Thistle Y. Harris

(a biography of Thistle Yolette Stead)

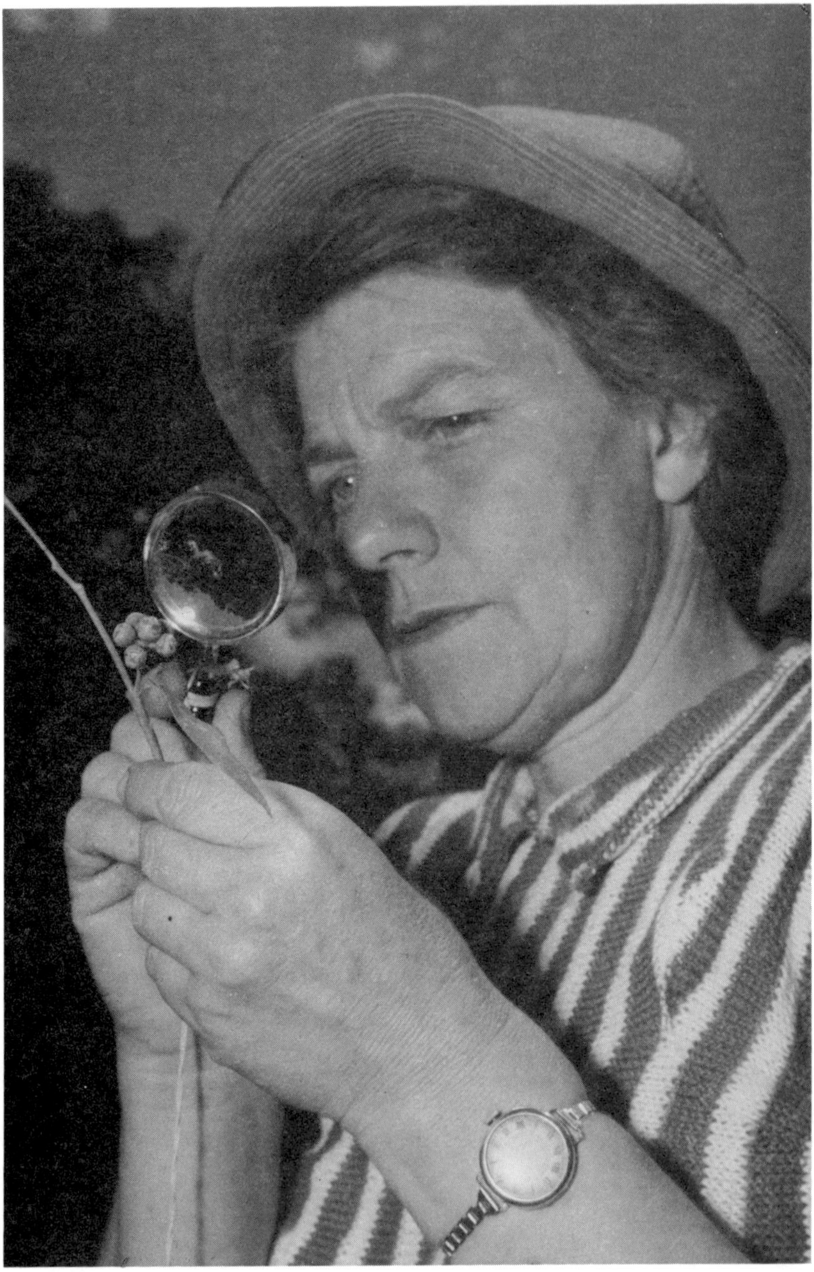

Thistle Harris — photograph published in People Magazine *on June 20, 1951.*

Thistle Y. Harris

(a biography of Thistle Yolette Stead)

by

Joan Webb

Published by

Surrey Beatty & Sons

The National Library of Australia
Cataloguing-in-Publication entry:

Webb, Joan B. (Joan Betty), 1929–
Thistle Y. Harris (a biography of Thistle Yolette Stead).

Bibliography.
ISBN 0 949324 85 X.

1. Harris, Thistle Y. (Thistle Yolette), 1902–1990.
2. Women botanists — Australia — Biography.
3. Botanists — Australia — Biography.
4. Women conservationists — Australia — Biography.
5. Conservationists — Australia — Biography.
I. Title.

580.92

Published November 1998

Printed in Australia by
SURREY BEATTY & SONS PTY LIMITED
43 Rickard Road, Chipping Norton, NSW 2170

Foreword

DR Thistle Stead, AM led a long, interesting, productive and varied life. She died in 1990 in her 88th year, having spent her life as a botanist, educator, writer, academic and advocate for the natural environment and other causes. Her energies were prodigious, her expertise renowned and at the same time, she was practical and generous.

She was initially a teacher. During this time she became actively involved in many conservation organizations. She was a member of the Wildlife Preservation Society, the Australian Forest League, the Junior Tree Wardens and many other organizations concerned with native flora and the environment. Her vision for Sydney as a garden city received wide publicity and her knowledge as a botanist enabled her to make recommendations for the planting of Australian native trees that would be appropriate to the narrow and often windy city streets. The work that she did in this regard in the 1930s should still command regard from our civic fathers. Much better that we have the red gums, lemon scented gums, paperbarks and New England peppermints favoured by Thistle Stead, than the harsh and somewhat alien palms that seem to be in vogue in Sydney at the present time.

As a lecturer at Sydney Teachers' College from 1938 to 1962 she was able to influence generations of student teachers in favour of the environment. During her time at the Teachers' College she wrote her first book, *Wildflowers of Australia* (1938) which made a great impression on those who were interested in Australian natural history. It confirmed her position as a leading exponent of Australian flora and the environment in which they grew. More books followed, each with a theme related to Australian plants, shrubs, trees and ferns showing the value of field work. She was generous in her recognition of the work of others in relation to the Australian native bushland, notable amongst whom were Joan and Eileen Bradley, those staunch supporters of conservation, authors and pioneers in bush regeneration.

Thistle Stead's interest in the Australian bushland led to her association with the National Trust of Australia (NSW). In 1962 the first section of the Wirrimbirra Sanctuary at Bargo was purchased to commemorate the work of David G. Stead, the long time companion and later husband of Thistle Stead. In 1963 the David G. Stead Memorial Wildlife Research Foundation of Australia commenced its association with the National Trust. In 1965 the Foundation suggested that the Trust take over the Sanctuary, with its 66 acres and improvements. This gift to the National Trust on behalf of the community was the first of the grand and generous gestures for which the community owes a great debt of thanks to Thistle Stead. She gave of her time, effort and money to improve the Sanctuary. By her Will she made a generous bequest to the National Trust, in recognition of which the National Trust has committed itself to the maintenance of the Wirrimbirra Sanctuary in perpetuity and to helping ensure that it remains a place for education and research — an appropriate memorial to David Stead and a tribute to a great Australian, Dr Thistle Stead.

This book by Dr Joan Webb traces Thistle Stead's life from her earliest to her final days. Her life, her idealism, her generosity, her professional career, her writings, her idiosyncrasies, her advocacy for the natural Australian environment and her recognition by an award in the Order of Australia are dealt with through the eyes of the historical researcher. Her relationship with David Stead and her involvement with many environmental causes, including Wirrimbirra Sanctuary and Kosciuszko National Park, are carefully documented in an eminently readable way.

Dr Webb has brought to her task of recording the achievements of Dr Thistle Stead her considerable capacity as an historian, a researcher and a writer. The National Trust of Australia (NSW) is proud to have helped ensure the publication of this volume so that the contribution of Dr Thistle Stead to her country, and especially to the natural environment of Australia, will be more widely known and appreciated.

The Hon. Barry O'Keefe, AM QC
President
National Trust of Australia (NSW) Sydney, 1998

Contents

Acknowledgements

FIRST I would like to express my appreciation to the many people who shared with me their memories of Thistle Harris Stead. It was a pleasure to meet so many interesting people who gave generously of their time and their personal anecdotes.

I especially wish to thank family members Carol Hinder and Alan Clark, and the Stead family members David D. Stead, Ivy Stead, Dorothy Basili and John K. Stead.

My appreciation is expressed to The National Trust of Australia (NSW) who provided financial assistance to enable the book to be published. Longtime member of the Trust, Ivor Wyatt, closely associated with Wirrimbirra for many years, was a great supporter of this biography.

My thanks to Alan Clark, Thistle's nephew, who gave me permission to use the Stead papers in the Mitchell Library, Sydney; to Alma Coombe at HarperCollins Publishers for assistance to view the Adam Forster paintings; to Professor Margaret Harris for permission to quote from the Christina Stead papers; to the National Library of Australia, Canberra, for permission to quote from the de Berg oral collection; to HarperCollins Publishers for permission to quote from their papers in the Mitchell Library, Sydney.

I wish to thank the publisher Mr Ivor E. Beatty of Surrey Beatty & Sons for agreeing to publish the book and for his help in arranging publishing details. Several people read the manuscript and offered constructive criticism and I thank Dr Tony Edmonds, Carol Hinder and Judith Steanes for this assistance. Ivor Wyatt read the Wirrimbirra chapters and Neville Gare helped with the final draft of the Kosciuszko chapter.

Finally I would like to acknowledge the help I received from the following people: Keith Armstrong, Dr Sid Baggs, Charles Boyd, Doreen Bull, Athol Berglund, Connie Burns, Elizabeth Chandler, Isa Cohen, Elaine and Betty Coldham, Tracey Crow, Peter Cuneo, Ross Doig, Joan Doney, Kath Elger, John English, Janet Fairley, Allan

Fox, Neville Gare, Ron Geering, Kay Geeves (Tas.), Roger Good, Eve Gourlay, Alan Gray (Tas.), Max Hewett, Dr Anna Hogg, Lawrie Johnson, Canon Alan Langdon, Eve and Ken Levis, Linda Lobendahm, Doreen Lyon, Patricia Macdonald, Dr Richard Mason, Dr Rex Meyer, Colin Miller, Brian Morley, Mary Reynolds, Shirley Rooke, Phyllis Schneider (Fackender), Vincent Serventy, Andrew Sourry, Judith Steanes, Shirley Stimson, Allen Strom, Ruth Sutton, Ian Tait, Pamela Taylor, John Teulon, Melva Truchanas (Tas.), Meg Walker (Tas.), Joyce Walker, Kathleen Wall, Peter Watts, Edith Winter, Vivienne Webb, Ivor Wyatt and Joyce Wylie.

Introduction

IT could be said that Thistle Harris Stead's life force with all its passions, projects and polemics took root when she was sixteen years of age. One crucial event gave direction and commitment to her life, dictated the career she was to follow, the books she was to write, the excursions she was to take and the contribution she was to make to conservation in Australia. Thistle Stead's consuming passion would often drive her to extreme lengths in confrontation with her contemporaries; it would engage all her energies until forced to give in to old age. What was this crucial event? A dedicated teacher from her school took the sixteen-year-old Thistle Harris to a meeting of the Naturalists' Society of New South Wales, and there she met, among other committed field naturalists, the self-taught naturalist, socialist and idealist, David George Stead, who at forty-one years of age was in the prime of his life.

Thus began a twin love affair, one with a man who was married to someone else for the first thirty-three years of their association, and the other with ideals and commitment to wildlife preservation that were firstly his, but appropriately, hers also. Thistle Harris was a botanist by profession and inclination, while David Stead was primarily a marine biologist, and here they differed, but their adherence to the ideals of peace, the need to save wildlife and its habitats, and a stand on atheism and socialism gave them a common direction in life.

Thistle became a well-known figure because of the books she wrote, but she was foremost an educator, and an outspoken advocate for the natural environment and its protection. In academic circles and the conservation bodies to which she belonged she was a forthright and upstanding figure who often succeeded in alienating those with whom she worked.

No one meeting Thistle Harris Stead would remain neutral; there could be anger, or annoyance, or acrimony, as well as respect, or affection, inspiration and admiration. This is her story, as well as, being in parenthesis, the story of environmental conservation in this country.

HARRIS FAMILY

Charles Thomas Harris m. Ilma Rokes

- Frances (Nan) m. Joe Clark
  - Alan
- Thistle m. David Stead
- Joan m. Peter Hunt
  - Carol and Wendy

STEAD FAMILY

David George Stead m. (i) Ellen Butters

- Christina

m. (ii) Ada Gibbins

- David Darwin m. (i) Percia (ii) Doris
- Catherine Ada (Kate)
- Frederick Huxley m. Mavis
- Doris Weeta m. Reg Kirton
- Samuel Kelvin m. Ivy
- Gilbert Jordan m. (i) Elsie (ii) Sylvia (iii) Betty

CHAPTER 1

An Idealist is Born — 1902–1924

"It is hard to appreciate what people or places or circumstances have most influenced one's life, but three individuals stand out very clearly in my memory: my father, one of my teachers, and my husband."

T. Y. Harris

"THISTLE! My only comment is that she should have been Rose". Such was the opinion of an English relative when informed of the birth of a second daughter to Charles Thomas Harris and his wife Ilma. What foresight to name this lively, quick-tempered child, Thistle, when she would grow up to advise her friends to pull out their roses and plant Australian natives. "Thistle by name, and Thistle by nature", her associates would say in later years.

"My mother was a person who liked fanciful names," Thistle Harris explained, "the name was given to me because I had a godmother, a stage actress, whose name was Thistle Anderson. My second name, Yolette, which nobody pronounces properly, I do not know where that comes from. She called my sister Frances Osra, the Osra, I think, coming from one of the novels of Anthony Hope Hawkins."[1]

Charles Thomas Harris had travelled to Australia from England, where he was born in 1874, in Kent, to accompany his father and step-mother, she being stricken with tuberculosis and recommended by the doctor to take a sea voyage. On arrival in 1890, they booked accommodation in Newtown, an inner suburb of Sydney, in a boarding house run by a widow, Eliza Rokes, who was also the mother of seven children, two of them still at school. Eliza was the great-great granddaughter of Thomas and Jane Rose who emigrated from England to the colony of New South Wales in 1792. Thomas and Jane were one of twelve farming families who were the first to volunteer to travel to Australia to assist the young colony to become self-sufficient. After an unsuccessful attempt at farming on the Parramatta River, Thomas and his family moved to the Hawkesbury River, where Rose Cottage is still open to the public.[2]

Eliza Rokes had been married to a farmer in the Ebenezer district, near Windsor, New South Wales, before coming to Sydney on the death of her husband. Her third youngest daughter, Ilma Richardson Rokes, was eventually to catch the eye of the young Charles Thomas. He was still trying to make up his mind about his future, one possibility being medicine, when he decided to stay in Australia while his parents returned to England. His step-mother died during the return voyage.

In 1891 Charles joined the clerical staff of the *Sydney Morning Herald* as a 17 year old office boy, rising eventually to become general manager in 1929, and remaining with the company until retirement. On December 11, 1897, Charles and Ilma were married at Moore Park; he was twenty-three and she was eighteen. Frances Osra (Nan) was born in 1898, and four years later Thistle Yolette was born on July 29, 1902. The family was complete in 1912 when another daughter, Joan Ellen was born, "specifically arranged for so that mother could have a companion." These three girls, through their mother, were fifth generation Australian.

The family lived in Mosman, first in Earl Street, until Thistle was seven years old, and then in Spencer Road for most of her younger days. It was a commodious house, with a large front verandah, and enough rooms for separate sleeping arrangements for the parents, already having difficulties by the time Thistle was in Primary school at Mosman Superior Public School.

"After the family meal, my sister washed up and I wiped. She had a basin over the sink, and she put things on a tray, and

I would dawdle. The rule was that the last person had to wipe the tray. Towards the end I would madly wipe and she would have to empty the water and wipe the tray — she never won on that. Also, Father had chops one morning for breakfast and steak the next morning, and the meat had to be bought every day, and I could never be found to go to the butcher's and my sister had to go."[3]

Thistle Yolette must have been an exasperating child, as she would deliberately hide in the garden on the other side of the huge coral tree, curled up on a rug with a book while her mother called, and Thistle would "just lie low". She was an avid reader; with half a crown Christmas money from her grandmother, she would buy Ethel Turner books until she graduated to her father's library, only a small collection of mainly fiction, but good quality. "He was a great reader of good fiction and we talked a lot; he introduced me to the classics, like Dickens, and the Forsyte Saga . . . I always offered to dust the library, but not much dusting got done — I'd read the books."[4]

Family outings at the week-end were an enjoyable affair at Balmoral Beach; travel by tram from Cremorne Junction, and tomato sandwiches for lunch. Christmas was celebrated in the traditional English fashion, at Father's insistence, while Mother grumbled about the midday roast meal, even though she was a good cook. Table manners at all times were most important, especially when visitors were entertained. One family with three daughters was particularly favoured by Father. "Mother was jealous of the woman. Father would play classical music which my mother hated also."

Thistle had a close relationship with her father, and enjoyed the stimulation of discussions with him; in her young teenage years she would sometimes accompany him to functions in the city. For childhood misdemeanours it was sufficient punishment just to have an "interview" with Father; he never resorted to physical punishment. Yet Thistle made the surprising admission in later years that "they didn't love me and at one stage I thought I must be adopted . . . I thought they didn't treat me well."[5]

An ability to keep up well with the work in primary school led this lively youngster to be something of a nuisance. Spurred on one day by another child in the class, Thistle busied herself one lunch hour by collecting all the pencils, rubbers and rulers off the desks and concealing them in the suitcases. Class resumed after lunch, no equipment — but teacher's suggestion to look in the suitcases revealed a jumble of implements, all to be sorted out. "At this stage I was just beginning to doubt about God — I was just beginning to make inquiries, and it was a Friday afternoon, the headmistress was informed but she was busy and said she'd attend to it on Monday. At home Mother informed me we were going away on Saturday for a holiday to the mountains for three weeks. So I decided there must be a God and that was reinforced when I got back to school and nothing was said."[6]

Aunty Thistle's Story

After Thistle's death in 1990, the family found a journal among her papers; in this journal she had started to write her life story, probably about 1984. However, the journal, in Thistle's handwriting, was almost illegible. Alan Clark, Thistle's nephew, asked his daughter Debbie to undertake the task of transcribing this fascinating insight into Thistle's memories of her schooldays. A few pages from this journal have been included in the present work to illustrate Thistle's eye for detail, her love of descriptive language, and her exuberant, fun-loving nature.

> In my primary schooldays, the long summer holiday had been for me the highlight of the year to be approached with joyful anticipation . . . But this summer it was different. I had just completed my primary school course and had left school with an admission notification, via the Qualifying Certificate Examination, to the local high school. As the high school served a large geographical area, containing many primary schools, competition was high. I was the only child from my school to gain admission that year. Now, instead of enjoying my days to the full and wishing they would never end, I was

counting the days when I would emerge from my primary chrysalis to a fully fledged high school acolyte . . .

Then came the major tragedy; just a week before the anticipated commencement of term I developed a mild, sniffling cold. Annoying, but, following the normal pattern lasting a few days. By the end of the week I was in bed, temperature well above normal, restless, unable to keep food down. Our doctor diagnosed influenza. "Nothing to worry about, Mrs Harris, there's a mild epidemic of it at present." . . . At last I was declared well enough to embark on my great adventure. In good time Mother and I sallied forth. I felt very important in my new blouse and skirt, my highly polished black shoes, and my hair plaited into two pigtails. I carried my new suitcase containing enough exercise books for use (one in each lesson that day), and my lunch packed in a container by Mother.

I was proud of my little mother. She was a good-looking, very dress-conscious young woman, early thirties, and judging by photographs a very attractive and vivacious girl (she married at 18). Her dress sense was hampered by lack of a reliable and regular income. My father, always generous, was also not a very practical man. I suspect too that he was drinking rather a lot, even at this time. Mother had to salvage what she could from her housekeeping allowance, which appeared to change with circumstances, to supply her clothing needs. She studied fashions on the well-laced women when she travelled on the ferry and tram, and paraded up Pitt Street [where] she studied the frock shop displays and made mental notes which she obviously retained well. She haunted leading city stores at sale time and raked through remnant [counters] collecting pieces of the correct length and pattern for her use. She was an expert seamstress with, in my opinion, very good taste, and always looked extremely well turned out.

First day in school: this day she was wearing a floral cotton frock in [various] overtones of blue, and a neat little plain dark blue jacket, a small black straw hat with a wreath of blue artificial flowers, a black eye-veil, black shoes, handbag and gloves. Very natty!

At the school we were told to wait outside the Headmistress' office, which we did, standing there rather formally for some twenty minutes while gowned figures went back and forth from a room obviously a staffroom, next to the Headmistress' office.

At last the door of the office opened and a gowned figure bade us enter. She herself left the room. We faced a large polished desk on which sat a telephone, a hand bell, a tray for papers, and behind which were bookshelves too far away from the flesh to be recognized from where we were standing. Seated behind the desk we saw the upper part of a large body, from the shoulders of which a gown fell over a mud-coloured garment. On a short neck reposed a large full face with blotched skin, a tight mouth, and little piggy eyes surrounded by puffy flesh. Heavy jowls wobbled as she moved, and grey hair was pulled back into a tight bun. I couldn't help comparing her with my natty little mother. Without speaking, she motioned mother to a chair but completely ignored my presence. I lowered my case to the ground and stood first on one foot, then the other throughout the interview. Still without speaking she drew a form from a drawer of her desk, took up the pen, and said, "Father's name?", and this continued until the entry form was completed. Then she looked up at Mother and said "That will be all". Mother stood up, saying, "Thank you" and walked towards the door. As she passed me I knew she would like to have stopped and given me an encouraging kiss, but the atmosphere forbade it, and all she did was give me a self-conscious smile and departed . . .

Dignified and gracious teachers glided silently along the wide corridors, dressed in academic robes, wrapped closely around them in winter and billowing like parachutes behind them in summer. They formed a continuous but interrupted

> procession (I had seen a formal academic procession once when Father took me to a conferring of degrees at Sydney University) . . .
>
> One [teacher] was large, very dignified, the only one who strode rather than glided along the corridors. Timid girls quaked at her entrance, though she was never known to administer punishment, or even reprimand. Her intimidating presence was adequate to maintain immediate and constant attention to her commands. She taught mathematics. Her introductory remark to the military like audience, [directed] to her standing class prefect, was "Sit the class". No greeting ever passed her lips. The class sat and work commenced, without a breathing space for forty minutes.
>
> Mademoiselle floated into the classroom, red hair piled in an untidy bun which threatened to, but never did, fall apart. She taught French by the then fashionable direct method: every word spoken by anyone was in French. . . It haunted me for more than seventy years that due to illness I missed the first week of school that year. So I never learned Mademoiselle's method of making contact with a first year group. What words did she greet them with? How did they respond? I asked my fellows, but none could ever tell me. Some looked confused and couldn't recollect, some just giggled, but all averred that No! M'selle never spoke in any other language. Then there was the delectable Latin mistress with wispy reddy-blonde hair and a bad limp. Benevolent, easy-going, giving and lovable Latin teacher. Her limp, we were told, was believed to be caused by hip disease which was incurable and she hadn't much longer to live. She had been engaged to be married before she suffered this misfortune, but her heartless fiancé had [broken] the contract. Such are the stories, I found later, that are ever rife in all secondary schools, always tragic, and rarely true.

Thistle quickly settled into her new school in 1915 — North Sydney Girls' High School — but she was an immature child, and minor misdemeanours, usually in the company of her boon companion Kathleen, soon led to regular detentions and frequent reports to the headmistress. The latter apparently made a complaint to the Department of Education, and a letter was sent to the parents before the end of the year. "We only giggled — surely the teachers could have coped with that." Father was very angry; he decided to send his recalcitrant daughter to Redlands private school at Cremorne, within walking distance of home, to follow in the footsteps of her older sister, Nan. Thistle was not informed of this decision until school was about to open in the new year, and she was never shown the report on her behaviour from North Sydney Girls' High School.

Unfortunately the pattern of restless behaviour continued at Redlands; she was far ahead of her fellow pupils, although records indicate she must have repeated first year because she was at the school for five years. Once again she was a regular visitor to the headmistress' office. "As the Intermediate Examination loomed ahead, I made what I thought were some very astute calculations. If, I reasoned, x girls sat for an examination from this school some must pass and, surely who among them but I? Henceforth I organized my life on the basis of this thesis. Imagine my chagrin when I came out with two language failures! My father, with whom by that time I was enjoying some intellectual companionship, was indignant at the poor judgement of the examiners. A re-marking of the papers brought one examiner to heel but the Latin man was made of more trusty fibre. So, a tutor had to be found to avoid another disaster."[7]

The unexpected failures in the Intermediate Examination of 1918 pulled Thistle up short, and the two years in senior high school saw a more conscientious application to studies. Moreover, it was an enjoyable time. Madame took the French class of seven, and worked them hard; there was no French name which was an equivalent of "Thistle" so in French classes our future botanist was called by the French word for "weed"! However, it was a teacher of English who became a major influence in Thistle's life, a talented woman who was instrumental in setting her young student on a path from which she would never stray.

Constance Le Plastrier was a wizard with words and had a profound knowledge of literature, her lessons being interspersed with whimsical tales concerning the lives of the classical writers. She was so erudite that she could substitute in almost any subject during an emergency; in Thistle's senior years she taught her, at one time or another, Latin, French, Mathematics, History, English and Botany. Constance Le Plastrier was engaged by Father Harris to tutor his daughter in Latin in after-school hours. But Botany was her first love and she was an excellent field botanist. "Companionably we sat round our big dining room table after school and diligently addressed ourselves to declensions and such like. But, however hard we tried, within minutes our discussions centred round botanical affairs. (I did manage a pass in Latin at the Leaving). Often she stayed on for dinner and drew my parents into more botanical talk."[8]

Constance Le Plastrier was the author of many study books in English literature for senior school students, often in association with a colleague, but most of these were published in the thirties, long after her association at school with Thistle. "The Life of Henry the Fifth" edited with introduction and notes by C. M. Le Plastrier and D. Brown, 1934; "A *Treasury of Verse*" (Selected and arranged by Edgar and Chilman). Companion notes and annotations by Constance M. Le Plastrier and Francis Jackson, 1932.

In 1916 Constance was a co-author with Agnes A. Brewster in the production of a radical new book, "*Botany for Australian Students*". This was a book on Australian plants for Australian schoolchildren, not only stressing descriptive botany, but stressing the importance of the ecology of Australian plants, an emphasis years ahead of its time. In the Authors' preface the need for the book is expressed: "The elementary text books available for pupils in the Australian schools are unsatisfactory, because Australian examples of botanical structures, etc., are absent. The object, therefore, of the present work is to supply this need. It is intended as an Elementary Text Book in which local types are used for illustrating the main features of plant life, so that the local bush may become the laboratory of the pupils . . ." Within the text, an interesting treatment of the structure and pollination of the grey spider flower *Grevillea buxifolia,* was the result of observations made by the authors themselves at Gordon and Mosman, suburbs of Sydney. The authors described their observations and their inferences, finally coming to a conclusion about the unusual method of pollination for this flower, just as accurate today in the closing years of the century.

Le Plastrier was a devout Roman Catholic and wrote several publications for the Australian Catholic Truth Society. One such product was a short story about a broken family and the joy of a return to the Sacraments of the Catholic Church — "*The Birthright*", 1913, 32 pages. A novel, "*Heirs in exile*", was published in 1910. It is interesting to speculate about Le Plastrier's failure to influence her charge in matters of religion when she had such a profound effect on her life in other ways.

As a member of the Naturalists' Society of New South Wales, and an active member of the Council of that body, Le Plastrier was one of the Society's most valued field botanists. One of her colleagues on the Council was David George Stead, who had encouraged her to join the Wild Life Preservation Society of Australia.

Sometime in 1918 there occurred a momentous event which was to change the lives of more than one or two persons. Constance Le Plastrier took her young pupil Thistle Harris to one of the Naturalists' meetings. Thistle met David George Stead and life would never be the same for either of them. She was 16 years old, and he was 41, the father of seven children. To say this was a "momentous event" is not an overstatement, or a lapse into the dramatic. Thistle never looked at another man; it was he who taught her the broad issues of conservation and its real value in relation to mankind; together they steered the Wildlife Preservation Society through many of its active years; and when he died she spent the energies of her later years in establishing his memorial, Wirrimbirra, the wildlife sanctuary near Bargo, New South Wales.

In later life Thistle would remember with pleasure the comradeship and bush lore she experienced in those early days with the Naturalists' Society. It was a personal kind of thing, she would say, not just fun-loving, but serious, too, with lectures, exhibits, discussions, and regular weekend excursions. Out into the bush they would go, with an enormous black billy from which they drank gallons of tea at afternoon tea-time, a group of people not so knowledgeable — Walter Froggatt, the Government entomologist, was an active exception — but all motivated to enjoy the natural world and learn more about it. For a young and impressionable girl, they provided an introduction to the exciting world of the natural environment, transferring to her something of their enthusiasm and delight.

At this time, too, Thistle's father encouraged her interest in natural history. "I can remember him promising, and then fulfilling his promise, of taking me out into the bush, and my first real contact with flannel flowers in the Sydney area was when we walked up Tumbledown Dick [Tumbledown Dick Hill at Terrey Hills], he puffing and panting because he was quite unused to this sort of thing, but really enjoying my enthusiasm, and this was a highlight of my school years, because I hadn't done this sort of thing before, and my companionship with my father was tremendously increased. I had a very good relationship with him but it was tremendously increased by this first — and only I may say — experience of his taking me into the bush. I hadn't really appreciated how extraordinarily beautiful the sandstone hills around Sydney were, not now but were then, and the flannel flowers were present there in enormous masses and made a great impression on me."[9]

Thistle realized her potential in that Leaving Certificate year of 1920, and was dux of Redlands School. She was awarded a prize for Botany, presented by Constance Le Plastrier, a volume by Robert Browning entitled "*Dramatis Personae and Dramatic Romances and Lyrics*." During this final year she realized she wanted to teach; after school she taught a child who lived nearby, one who didn't attend school, and with the money earned she bought more books. Through Constance Le Plastrier she met Hilda Butler, another young enthusiastic naturalist, who was coached by Le Plastrier and also introduced by her to the Naturalists' Society. Hilda and Thistle visited Sydney University several times during their Leaving Certificate year of 1920. Success for Thistle in the Leaving Certificate led to the award of a scholarship which enabled her to enrol at the University in the Faculty of Science. Her mother was less than enthusiastic about this turn of events, believing her daughter should be earning money to help the family, but her father's support was a deciding factor.

The only subject in which Thistle was really interested was Botany, but in First Year Science was forced to take three other subjects. She was strongly advised by the authorities to take Mathematics, which was not her area of strength — "Miss Le Plastrier taught me mathematics . . . and while she was an extraordinary person, she wasn't a good teacher, I think, and my mathematics were pretty poor." But she enrolled in Mathematics, and found herself completely confused. Fortunately, Zoology classes didn't start until second term, so she arranged a transfer, to the mutual relief of both herself and the professor of mathematics.

Life in the Botany school provided deep satisfaction and a great deal of fun. The Botany Department was staffed by three Scottish lecturers: Professor Lawson, D.Sc. (Glasgow), had been appointed in 1913; John McLuckie, also a Glasgow graduate, had been at Sydney since 1915, but Patrick Brough, another Glaswegian, had only arrived in 1919.

In class and on field trips, Hilda and Thistle harried these men unmercifully, inexperienced as they were in Australian taxonomy. "[We were] armed with precious copies of Flora of NSW ('rare and much sought after' as was written in the copy my father persuaded a friend to disinter from the archives of the Lands Department) and the knowledge of classification the wise naturalists had given us."[10] Paddy Brough had a hard time as the girls added insult to injury by initiating him into cooking chops over an open fire and denying him the tomato sauce and cutlery which he timidly demanded.

Thistle and Hilda took a childish delight in taking specimens of plants for identification to their Scottish lecturers ("extraordinarily good biologists, but very little knowledge of the Australian flora"), who took all sorts of side-tracks to try to arrive at a correct identification. In the Botany Department at that time there was a very capable man who held the position of senior attendant; he had an extensive knowledge of field botany, and would later become one of the authors of "*Flora of the Sydney Region*".[11] This man was Obed David Evans. To him the girls would go to check their identifications before going to the Scottish lecturers, and to him the Scotsmen would go to make sure they knew the answer. "This went on throughout our University life, so that we didn't really make ourselves very popular in the Botany Department, but we had a very good time." In 1954 Dr John McLuckie, in association with Dr H. S. McKee, published a comprehensive and well presented textbook for university students, "*Australian and New Zealand Botany*". In the note about the authors, attention was drawn to McLuckie's papers on the ecology of Mount Wilson and of the Kosciuszko Plateau, pioneer Australian contributions to this important branch of Botany. The book was dedicated to A. A. Lawson, first Professor of Botany in the University of Sydney, "a generous and helpful colleague, a stimulating teacher, a brilliant botanical investigator." Perhaps Thistle was short-sighted in her youthful criticisms of her Scottish instructors.

In spite of a reputation for lively clowning and practical jokes, Thistle did carry the blame once (and this time unjustifiably) for writing under a notice about an excursion which detailed "drinking vessel" among requirements, that "an ordinary coop or moog will not do". It was actually the work of Pat (P.D.F.) Murray, later Professor of Zoology in the University of Sydney.

A more serious incident (trivial by today's standards) nearly led to Hilda and Thistle being dismissed from University in First Year. The girls were anxious to visit a flower show scheduled for a Tuesday morning following a Monday holiday in October; sometimes the Tuesday morning lecture was cancelled, and the girls gambled on this being so and went to the show. Later John McLuckie confronted Thistle, "You weren't at the lecture," "No, we couldn't make it." But another student had written their names on the attendance list. Thistle went to the attendant and requested their names to be crossed off. The matter was taken to the Dean, Edgeworth David (Geology), who showed a deal of concern; the culprits were allowed to attend lectures while the case was being considered. "We were very juvenile compared with 18-year olds to-day." Apparently the outcome was a favourable one for the two aspiring scientists.

These were hard days financially for students on scholarship — a pie and two plates at Grace Brothers on Broadway had to suffice for a midday meal; gingham at threepence a yard made useful blouses and skirts. Girl botany students on field trips in the bush had to wear jodhpurs — slacks and shorts were unknown, and there was no question of not wearing hats. Hilda came from a poor Irish Catholic family, seven children, and not a happy home. Her friendship with Thistle must have been a great comfort to her; the girls did not attend many University affairs, content to share in the social life of the Naturalists' Society, or go out at weekends as a pair. Botany students at the University were expected to collect and identify several hundred specimens of Australian native plants, pressing the plants to make a personal herbarium. Hilda and Thistle did a lot of collecting together, but since Thistle lived in Mosman, most of her collecting was done in the area around the Spit and Seaforth, both places now heavily built up. Not long after she graduated, Hilda entered a closed order of nuns and died quite young.

Thistle graduated as Bachelor of Science on Thursday, May 29, 1924. Her grades each year had been acceptable, with more than satisfactory results in Botany. First year (1921) — pass in Chemistry, Physics, Zoology, High Distinction in Botany; Second Year (1922) — Distinction in Botany, Credit in Agricultural Chemistry; Third Year (1923) — Distinction in Botany, Pass in Agricultural Chemistry. Father Harris was proud of his daughter, but Mother was jealous of the relationship between them; she refused to attend the graduation ceremony. By this time the parents were quite estranged, but Charles did not help the situation by encouraging Thistle to go to his room in the evening to discuss what each one had been reading during the day.

On completion of her Bachelor's degree course, Thistle had rosy dreams of continuing her botanical studies in an honours year, but finance would not permit her to enrol as a paying student. She approached Professor Lawson to ask for a job as a demonstrator. "He was so horrified at the thought of a woman in the job that he could only say er . . . er . . . and show me to the door. Looking back I surely did him an injustice — he probably feared my antics in the lab."

Next she tried for a position in the Sydney Herbarium, this time convinced her failure to secure a place was due to her being a female. Unable to continue her botanical studies, Thistle spent 1924 as a student at Sydney Teachers' College, gaining her Diploma of Education. This was not a last resort because with her choice of teaching as a career there was a good deal of idealism involved.

The Principal of Sydney Teachers' College in 1924 was Alexander Mackie, who had been in that position since 1906, and remained there until 1940. At this time, 1924, the new College building in the University grounds was still incomplete; students and staff had to divide their time between the new building and Hereford House, Glebe, while some work was still carried on at Blackfriars, where the library was located. Thistle was in her first year of teaching in a high school when the building in which she was to spend the later years of her career was officially opened on Monday, March 23, 1925, by Albert Bruntnell, the Minister of Education. In 1921 Mackie had expressed misgivings about the design of the new college, asking for a "large, central quadrangle instead of two cramped courtyards"; but his request was rejected, and the stage was set for two courtyards, one for men and one for women, a segregation which existed for decades.

One of Thistle's lecturers would have been Gilbert Phillips, a former teacher in Victoria, who held a Master's degree from Melbourne University and a D.Sc., from London University. In 1920 and 1921, one of his students had been Christina Stead, eldest daughter of David Stead, forbidden by her father to attend University and reluctantly preparing for a career as a primary teacher. Gilbert Phillips had observed Christina's interest in psychology, and in 1922 offered her a position as his research assistant in psychology. This unfortunately was only for one year, and after an unsuccessful year as a classroom teacher, Christina was once again rescued by Gilbert Phillips who employed her at the Teachers' College as a Junior Lecturer in psychology. This was 1924. Thistle and Christina must have rubbed shoulders now and then at the Teachers' College, but no confirmation of their meeting is available. Phillips became Vice-Principal of the College in 1944 and replaced C. R. McRae as Principal four years later.

In 1921, 360 College students were attending the University, and in 1924 the Diploma of Education, although a University award, was the teaching responsibility of the College staff. Enrolments grew from 56 in 1922, when Economics graduates were admitted for the first time, to 60 in 1925.

Both staff and students treated Alexander Mackie with great respect and affection, and although his views were not always accepted, his authority was never seriously challenged. He was not autocratic, held regular staff meetings, and listened to the views of others. "He was an outspoken Scot who lacked finesse and who believed in being confrontationist where necessary. Under his guidance the College became a major force in the educational system of New South Wales and one of the leading institutions of its kind in Australia."[12]

In 1924 at Sydney Teachers' College, Thistle was Honorary Secretary of the College branch of the Naturalists' Society, the branch having been established in September, 1923, and in 1924 boasting of some ninety enthusiastic members. The group held two meetings in each of the three terms, one excursion in term 1, and two excursions in each of terms 2 and 3. "On Monday,

May 5, Mr D. G. Stead delivered a lecture, entitled 'By Sea and Jungle in the Malay Peninsula', speaking of the people of Malaya and of their habits; he also made some interesting remarks of the fauna and flora of this region." So ran part of Thistle's report on the Naturalists' Society in the College magazine, *The Kookaburra*, December 1924.

As Thistle's days as a student drew to a close, her thoughts turned to her future career as a teacher of biology in a secondary school. Was this the ideal situation for someone of her training and enthusiasm?

"Being a graduate student I entertained, as was common among us, a tolerant contempt for non-graduate students training for appointments in primary schools. At this time one-teacher schools were usual in country districts, often remote, and my tender academic heart bled for the poor unfortunate pupils who would be subjected to the unskilled teaching of the short-term trainees. How unfair, I reasoned, that these already under-privileged children should be denied the benefits of really skilled teaching. So I sought an interview with the Director and nobly offered my great skill and knowledge for a one-teacher school. He promptly refused my offer, pointing out that, with my training, I would command a higher salary than the Department could afford to pay for the post I wanted. Apparently there was no book keeping machinery to adjust that. Since then I have often wished I could have heard this Director entertaining his colleagues with the story of this misguided and arrogant idealist."[13]

Chapter 1 Notes

1. De Berg Tapes, National Library, Side 2, p. 8.
2. Information supplied by Carol Hinder.
3. Mitchell Library Tape, Stead Papers, Box 3(25) MSS 5715.
4. ibid.
5. ibid.

6. ibid.
7. Magazine article in *Community*, August 1995.
8. ibid.
9. *Good Gardening,* October 1980.
10. *Community,* August 1995.
11. "*Flora of the Sydney Region*", N. C. W. Beadle, O. D. Evans, R. C. Carolin, first published in 1963.
12. P. 85, History of Sydney Teachers' College.
13. *Community,* August 1995.

CHAPTER 2

An Unusually Gifted Man

"I thought he was a mountain of a man."

Allen Strom

DAVID George Stead, who was to become the husband of Thistle Harris in 1951, was one of Australia's most talented pioneer naturalists and conservationists, a self-educated marine biologist and naturalist who, it is claimed by some, was largely responsible for the ideals and philosophy behind our environment movement today.[1] He was born on March 6, 1877, sixth child of English-born Samuel Leonard Stead and his wife Christina McLaren, who had come to Australia from Scotland as a baby. Samuel, a builder and carpenter, had purchased land in Walker Street, not far from the harbour in St Leonards (now North Sydney), where he built a modest two-storey weatherboard home. There Christina produced seven children by the time she was 32. One daughter, May, died in infancy, leaving three boys and three girls: Sydney, Samuel, Christina, Jessie, David and Florence.[2]

David's mother was nonconformist in religion, and strict, with many taboos; no dancing, smoking, card playing, alcoholic drinks, theatre, and so on. She was a member of the Plymouth Brethren, and the younger Stead children, David and Florence, sometimes accompanied their mother to meetings. Samuel, on the other hand, had "habits" about which the women spoke in lowered voices: he took snuff, and he liked a glass of port. Furthermore, he was a freethinker, as atheists were then called, a member of the Order of Odd Fellows (of which he became Grandmaster), and belonged to the Dickens Lodge, where members recited and acted from the famous Dickens tales.[3] But it was a happy family, a musical family, whose members would gather around the piano and sing the old favourites.

After a local public school education, David was apprenticed at age twelve to the Sydney rubber-stamp maker Karl Faulk, where he learned the art of lettering that was his pride, and which he was to use to good effect in his later position as a junior in the Fisheries Department. He went to Sydney Technical College to study zoology and during the course boiled down and mounted a cat and a dog, which travelled with him in glass cases into all his future homes. But in the August of 1893, when David was only 16, his mother died, making him promise to keep her rules of life, and this he did. He became a puritan like his mother, never indulged in alcohol, or the theatre, or dancing; but he was an atheist like his father, and inherited his good looks and lively nature. His fair hair and blue eyes were to attract attention from admirers well into his manhood.

David was a passionate young man, and his passions found expression in his enthusiasm for natural history. He joined the New South Wales Naturalists' Society, a group of committed amateurs, and in 1898, at 21, became a member of the professional scientific body, the Linnean Society of New South Wales. By 1900 he had several short articles published by the prestigious British Linnean Society.[4] Later, in 1910, when elected a Fellow of the Linnean Society in London, he was said to be the youngest person ever elected to that favoured position. In 1901, following his writings on international affairs, he was made a member of International Conciliation, one of the first world peace bodies, based in New York.[5]

Florence Stead, David's younger sister, was working at David Jones, the Department store, as a seamstress, and introduced David to her fellow worker, Ellen Butters. David and Ellen were married on August 17, 1901, but it was a difficult time for Ellen, who came from a very devout family. The 24-year-old groom refused to take part in a religious service and the bride's father refused to welcome the married couple into the family home.

By 1902 David's expertise in the area of marine studies led to his being appointed a scientific assistant under Harold Dannevig, Director, in the NSW Fisheries Commission, employment being from May 1902 at the annual salary of £200. His home in Rockdale boasted a menagerie in the back garden, where he kept venomous black snakes, and a diamond snake. Two months after his appointment to the Commission, a daughter was born to Ellen, on July 17, 1902, twelve days before Thistle Harris was born only a few suburbs away; she was named Christina Ellen Stead. Grandfather Butters refused to speak to his daughter or look at his grandchild.[6] Three years after her marriage Ellen died, her family claimed of a broken heart, although her death certificate records a burst appendix followed by septic absorption and heart failure eighteen days later. She was twenty-eight when she died in Prince Alfred Hospital, Camperdown, on December 9, 1904.[7]

The girl-child Christina was mothered by her father, who became a dominating figure in her life, with far-reaching effects on her personality. He told her stories peopled with the sea monsters and land animals he loved so much. His youngest sister, Florence, went to live with Christina and her father immediately after the death of Ellen, keeping house and looking after Christina and her own daughter, Gwen, who had been born in July 1903. Christina wrote later in life, "these were very gay years with the brother and sister in the house together; but David was very young, he had to remarry".[8]

David Stead met Ada Gibbins through her father, Frederick Gibbins, a man of some means, with business interests in oyster farming and timber. Ada was a dark, pretty young woman, but of fragile health; she had attended a private finishing school in the Blue Mountains where she learned ladylike accomplishments, and at age 27, when she met David Stead, was living at home with her parents and one unmarried brother, wondering about her prospects. It seems, from the beginning, to have been a marriage more of convenience than of love.[9] Christina would always claim that her stepmother was married off by her father, Frederick Gibbins, against her will.[10] The wedding was held on New Year's Day, 1907; David Stead agreed to a religious service, but no church, so the Anglican rites were performed in the drawing room of the Gibbins home, "Dappeto". Frederick Gibbins offered the couple a big house, rent free, on Lydham Hill at Bexley, and there David and Ada Stead lived for the next ten years, producing a family of four boys and two girls. The boys were named after scientists: David Darwin (born 1907), Frederick Huxley (born 1910), Samuel Kelvin (born 1915) and Gilbert Jordan (born 1917); Catherine Ada (Kate or Kit) was born in 1908 and Doris Weeta in 1912.

Christina wrote in later life, "When the father remarried, I was past four, and soon I was looking after a younger child and then younger and then I became the cradle rocker and message runner and the one who sang the sleep and told tales".[11] She was to write also in another context, "My stepmother did not like me, very natural, as I was the kind of child only a mother could love and then probably with doubts: her treatment of me was dubious".[12]

In January 1917 Frederick Gibbins died, and to the family's dismay the houses at Dappeto and Lydham Hill had to be sold to pay off his debts. David Stead found a house at Watson's Bay, a two-storey weatherboard building at 14 Pacific Street, very much in need of renovation, but close to the harbour and a delight for the children. He named the house "Boongarre", after an Aboriginal chief who had lived in the area. But life was not always sunny; there was insufficient income to support a family of seven children and maintain the house in reasonable condition. Ada, bitter about the wretchedness of her situation, rarely even spoke to her husband. Christina wrote dramatically about this period in the family life in her book "*The Man who Loved Children*". Sam, the father, was based on her father, David, and Henny was his second wife, step-mother to Sam's daughter, Louie. "Louie had actually once or twice had moments when she could listen to Henny's scoldings and (although she trembled and cried bitterly) could recognize that they came from some illness, her neuralgias, or cold hands and feet, or the accumulation of bills, or from

Sam's noisy joys with the children, and perennial humanitarian orations."[13]

In 1941 Thistle Harris was to receive a parcel from New York from Christina Stead. It contained a copy of Christina's book "*The Man Who Loved Children*" and contained the following inscription:

To Dear Thistle —
A Strindberg Family Robinson —
in some respects might be considered a private letter to Thistle
from
Christina Stead
New York August 1941[14]

In April 1942 Christina wrote to Thistle saying, among other things, "Yes, I feel that the characters in TMWLC are very very real . . . I am always surprised at my memory for detail . . . Sam did really boil a fish for oil . . . and Henny really did play cards out of desperation and so on, so on: it did not happen exactly all on one day but what I did is a very real representation of life as it was then. As to the details: I am very happy to hear you say that you recognize them. I did not invent, naturally . . ."[15]

Henny, the dispirited wife, says to her husband, Sam: "I've had the house stinking like a corpse cellar with your formalin that you're proud of and had to put up with your vile animals and idiotic collections and your blood-and-bone in the garden and everlasting talk, talk, talk, talk, talk", she screamed in a hoarse voice, "boring me, filling my ears with talk, jaw, jaw, till I thought the only way was to kill myself to escape you and your world of big bluffs and big sticks, saving the whole rotten world with your talk".[16]

In her old age, Thistle Harris would say that she saw *The Man Who Loved Children* as "substantially true", but she also thought the novel "tremendously cruel".[17]

In spite of possible shortcomings as both a husband and a father, David's scientific work surged ahead. He was a prolific writer; in 1904, using the pen-name "Physalia", he wrote for a New South Wales newspaper *The St George Call* a series of 50 weekly articles under the title "Nature Notes". His published articles, papers and reports — most of them issued by the New South Wales Government and various natural history and scientific societies in Australia and beyond — comprehensively covered the wide field of Fisheries, Fishes, Crustaceans and allied matters. His best-known books and papers are *Fishes of Australia* (1906); *The Edible Fishes of NSW* (1908) — said by Gilbert P. Whitley (Fisheries Newsletter, September 1957) to be "still most readable and authoritative to this day"; *Crustaceans, Ancient, Modern and Mythical* (1905); *Additions to the Fish Fauna on NSW* (1907); *Eggs and Breeding Habits of Fishes* (1910); *Fisheries of NSW* (1910); *The Future of Commercial Marine Fishing in NSW* (1911); *Whales and Whaling in Australian Seas* (1927); *The Tree Book* (1933); *Giants and Pigmies of the Deep* (1933); *The Rabbit in Australia* (1935); *The Fisheries of British Malaya* (1923); *Marine and Fresh Water Fisheries of Australia* (1924). He left behind a completed manuscript *Sharks and Rays of Australian Seas*, published posthumously by Angus and Robertson in 1963.

Stead was always concerned to translate biological detail into language that non-experts could understand. His best known published works *Fishes of Australia* (1906) and *Edible Fishes of NSW* (1908), were also acclaimed for their detailed photographic plates and drawings. The former Curator of Fishes at the Australian Museum, John Paxton, said Stead's contribution as a fisheries biologist was significant. "His *Fishes of Australia* is a popularization. He, throughout his career, I think, attempted to popularize science, to educate the general public on those things that he was interested in, and certainly in his *Fishes of Australia* this is a book that wasn't written as much for the scientific community as for the general public." Yet Paxton agrees that the photographic plates in his *Fishes of Australia* are of a quality that scientists could appreciate. "He did a number of drawings himself and his drawings are quite good, but with his photographs he took the time to preserve the fish, to have all the fin rays standing up, and from a scientific point of view, you can, from his plates, count the number of scale rows the fish has, count the number of fin spines and rays . . . today we use colour, which was not available to him at that time. So his illustrations are excellent."[18]

In 1914, as Special Fisheries Commissioner, Stead travelled overseas to investigate British, European and American Fisheries for the New South Wales Government, returning to Australia in 1915. His task in England was also to buy three modern steam trawlers in Hull. "The trawlers sailed out across the world and for five lively years we had nothing but trawling and fishing talk."[19] Stead was to be General Manager of the State Trawling Industry from 1915 to 1920, and this setting up of Australia's first state trawling industry was probably his most radical achievement. It was a visionary socialistic scheme to harness cheap food for the masses, and not only did he have to campaign to sell the idea, but he was responsible for setting up the organizational machinery to make the industry viable. A pamphlet published in 1915 shows his enthusiasm:

> "Cheap food for the people. What state fish costs. State trawled fish is as choice as it is cheap. Inshore and estuarine fish is also obtained direct from the fisherman and retailed at prices from one-third to one-half of those ruling in the establishments of grasping middle men."[20]

However, for several years during his post as manager of the State Trawling Industry, he had been the target of defamatory attacks by the conservative Opposition in Parliament. He was accused of "fishing out" the waters, of costly mismanagement; in 1916, the first of a series of damning government reports was produced calling Stead's administrative competence into question. By mid-1918 allegations that tons of fish had been condemned by health inspectors and dumped at sea were widespread; the previous year's loss in the state trawling industry was put at twenty-four thousand pounds.[21]

A Royal Commission into the State Trawling Industry conducted by Mr G. Mason Allard, was set up in April 1919, and the pressure on the government continued. In his report of

January 1920, the Commissioner expressed his appreciation of Mr Stead, "The amount of labour, the scientific knowledge, the power of acquiring technical and practical knowledge, the enthusiasm and love of the work which [he] has brought to bear upon the industry from its inception, even in most disheartening circumstances, and he deserves considerable credit for his efforts". The Commissioner went on to say that Stead was first and foremost a scientific man and "has not throughout exhibited the possession of the business sense".[22] Mr Allard recommended that Stead's "undoubted attainments and scientific knowledge concerning fish and fish culture should be more effectively used in connection with the work of the Fisheries Department throughout the country than has been done in the past, for in some measure these qualifications have been wasted of recent years . . . I suggest that there should be a Board of Advice to the Minister . . . Mr Stead to be consultant to the board; the board to meet once a week; and the members and consultant to be suitably remunerated".[23] Stead was subsequently relieved of his post as general manager of state trawling early in 1920.

David Stead threw himself into work for the natural history societies, in particular the Wild Life Preservation Society of Australia, of which he was the co-founder in the year 1909. From that time until his death in 1957 he was an office-bearer, holding at various times the position of President, Vice-President, Honorary Secretary, and at the time of his death, Editor. But of those people who supported the venture, it was David Stead who implemented the enterprise, actively promoting it to the end of his life.

Some years prior to the establishment of the Wild Life Preservation Society in 1909, David Stead and others had discussed among themselves the urgent necessity for the formation of a special organization which could crystallize in its constitution and activities the growing interest and concern regarding the future of Australia's native wild life. Stead reported[24] that the first practical move to that end was actually made by a representative of Sweden — his friend, Count Carl Birger Mörner, a well-known naturalist and poet, who was, at that time, the Consul for Sweden in Australia. Count Mörner had particularly sought that Consular position so that he might the more readily follow up his studies of the Australian fauna, serving in the position from 1906 to 1910. Stead had discussed the formation of such a body before the Naturalists' Society of New South Wales in 1901, and had published items relating to this in the columns of the daily papers — notably the *Sydney Morning Herald* in 1900, and February 1902, but the suggestion bore no fruit.

Following the canvassing of the idea of the new preservation body among a number of naturalists, nature lovers, and some interested public figures, a preliminary discussion meeting was held at the offices of the Royal Swedish Consulate on May 11, 1909. The outcome of this meeting was the decision to call a public meeting for the formation of the Society. This was held in the Royal Society's Hall, 5 Elizabeth Street, Sydney, on the night of May 19, 1909. Despite the inclement weather, fifty people attended, fifty members were enrolled, and within a week, this number had been raised to over one hundred. The first President was the Honourable F. E. Winchcombe, M.L.C., who, apart from his public office, was the head of the well-known firm of Winchcombe, Carson and Co., woolbrokers, skin and hide merchants. David Stead was one of four Vice-Presidents.

In 1956 a book, *Naturecraft in Australia*, was produced by the Wild Life Preservation Society of Australia, edited and arranged by Thistle Y. Harris. An appendix outlined briefly the charter of the Society, pointing out that it existed for the purpose of encouraging the protection and general conservation of Australian fauna and flora. "The general policy of the society through the years has been an educational one; for its officers believe — and always have believed — that a conservation-minded public is the best safeguard the bushland can have. It aims, therefore, at educating the people to understand the desirability of a conservation policy, through a knowledge of fauna and flora and its environment . . . In addition, it opens the gate to an understanding of the relationships between the plants and animals living in the bush, and of the factors necessary to their continued existence."

The protection of all marsupials was part of the Society's early work; the saving of the koala, in particular, was a long fight. Skins labelled "wombat" were being exported to the United States, and when David Stead discovered this, he wrote to the United States President Herbert C. Hoover. In response, Hoover agreed to ban all skins, however labelled, and the koala trade, robbed of its most important market, soon ceased. This was a major victory.

Today the work of the Wild Life Preservation Society of Australia continues under the capable leadership of President Vincent Serventy. In more recent times the Society has played a leading role in the reservation of a long list of national parks, and in the creation of many World Heritage Areas. The conservation movement today owes a debt to founders like David Stead, and the following tribute, expressed at the time of his death, was one of many praising his idealism and enthusiasm. "It was a privilege to have been associated with him in his wonderful record of work for the conservation of our Flora and Fauna. The status of the W.L.P.S. and its reputation today are a monument to him as, for very many years, he was the power and his the initiative that inspired its work. The World — and Australia in particular — is the better for one David G. Stead having passed through it."[25]

Allen Strom, who was Secretary of the Wild Life Preservation Society immediately after Stead in the 1940s, said Stead was crucial in getting Governments to take the conservation movement seriously in the early years of this century. "I though he was a mountain of a man as it were, in other words, he dominated the scene. He demonstrated in those times that I first met him that he was dedicated to the cause of what they now call 'Nature Conservation', that was called Wild Life Preservation in those days, and I was quite impressed. The point was, he'd been fundamental in the establishment of the organization back in 1909, and by the time

I got there of course, it was a number of years later, over thirty years later, and the organization was well established, it was well known, and of course it was effective in the community, as effective as you might imagine any organization at that time would be. After all, let us not forget that there was very little in the way of nature conservation from the point of view of conviction of administration by the '40s. In other words, the whole thing was just beginning, and I think that all that was due entirely to the perseverance of this man, and the way in which he could influence other people to be in it as well."[26]

But Stead's lack of an academic education was to him a bitter disappointment all his life. As a young man active in the Naturalists' Society and the Linnean Society, he found himself brushed aside by the academics, not prepared to listen to a young man with no academic qualifications whatsoever. "He was conscious that he was a nobody as far as the academics were concerned, and it acted as a frustration to him all his life and I think it affected his whole personality." So said Thistle Harris in an interview in 1987.

Among his friends in the Naturalists' Society of New South Wales, David Stead did not feel inferior, because few of them were academics, although possessing profound knowledge in their own fields. The unwritten code of behaviour among the Naturalists was, if you have knowledge, pass it on and share it around. Many of the members were David's personal friends — people he met in trams, on buses, on the beach; in talking with them he aroused their interest.

David Stead was a great lover of Australia; in his library there were Henry Lawson and Banjo Paterson. *On Our Selection* was there, but Henry Handel Richardson, for example, was far too literary for him, Christina Stead recalls.[27] There were no novels in the library — he didn't approve of novels — but there were of course books on science, history, and psychology.

In 1919 David Stead told his daughter Christina that he had fallen in love with a schoolgirl her own age. Christina, who often went on the Naturalists' excursions and knew Thistle

Harris was greatly shocked, and it seems that David himself decided the romance should not develop beyond a friendship of kindred interests.

Apparently still in the employ of the Public Service, David Stead was offered a two-year posting as Fisheries Commissioner and Director of Food Supplies in British Malaya. He left Sydney on December 31, 1921, bearing with him a letter from the Lord Mayor; its contents strike a euphemistic note after the calumny and criticism of the previous years:

> "The bearer of this document, Mr David G. Stead, is one of the leading public officers of the State of New South Wales and is about to commence a travel on official service to British Malaya, his services having been made available to the British Government there by the Government of New South Wales for the purpose of reporting upon Malayan Fisheries.
>
> "Mr Stead is held in high esteem in this country both for his services to the State in various official capacities for close on twenty years and for his efforts to preserve the Australian fauna and flora.
>
> "He holds a somewhat unique position in Australia. In this connection his efforts have been recognized by the principal world authorities both because of his published works and his activity in deep-sea fishery development."[28]

The letter goes on to list his involvement in the various professional and amateur societies to which he belonged.

Stead was a rebel in Malaya; he was shocked at living conditions there and made his views known in the local press. He believed in equality; when in Singapore and invited to sit on the platform by the Governor of the Federated Malay States he refused because there were to be there no native representatives, only British people.[29] His physical health suffered too, and he was lonely and homesick; unlike his white associates he never allowed himself the relief of alcohol. He wrote to the family often, continuing to write in the nonsense language he had always employed for the children's entertainment. A letter dated May 27, 1922 is written in this vein:

> "I expect you arr trieing too B a fairlie goode boie if you aint you betr start now so you will no the way wen I kum home. Luv to evryboddie and all be goode. Thoes who duz thare dutie ar luvly wuns they will get some frootie and some kurren bunz. From DadPad the Boald."[30]

There is no direct evidence that David Stead and Thistle corresponded when he was in Malaya, but Christina Stead, in *The Man Who Loved Children*, writes of Sam Pollit's correspondence with the young naturalist Gillian Roebuck. Christina was David's confidante on most of his personal affairs so it is not unreasonable to conjecture that she knew of his contact by letter with the young naturalist, Thistle. Gillian concludes her letter to Sam, "Yes, I am serious about Wild Life: it gives me a wonderful feeling for nature and has expanded my interests: I really love nature now, thanks to your teaching: I mean real love. It was just something to do before".[31] Sam returns home from Malaya and the family move house after the death of Henny's father. "Spring was coming on and Sam was very restless. For weeks he would love Gillian Roebuck; then he would go to see Saul Pilgrim's sister, Mrs Virginia Prescott, a widow . . . Sam did not love her, but when his feeling for the nature-spelled girl, Gillian, became too strong, he went and talked to Virginia. He was unable to see Gillian because they both felt they were too conspicuous in either Baltimore or Washington, and Sam despised hole-in-the-corner meetings: it was not worthy of them."[32]

Before he left for Malaya in 1921, David Stead had commenced a weekly children's feature article which appeared as "The Great Outdoors" in the *Sunday Times*, using the pen-name "Dinnawan." He continued writing from abroad until the last article appeared in April 1923.

> "I am Dinnawan the emu, and though I must confess that I am especially fond of the great grassy plains and the glorious

rolling downs, I am really not so very particular about the kind of country so long as there are some nice open spaces here and there. But wherever I am, and for all my life, I've been filled always with the desire to know, about everything that I've seen, just like you, Everyboy. Folks do call me inquisitive, but then, I've heard them say that about you, too. Well, Everyboy, here we meet again, and I hope you've brought a pair of well-oiled legs with you this morning, for we're going on a long bushwalk . . . I've brought for company Dirra-diree the wagtail, and Googoo-gaga the merry jester, Old Jack. Dirra-diree will spy out the land for us, and when he is tired, he can rest on my head for a while."[33]

As something of a compulsive communicator, David Stead also used radio to educate children. He spoke of his travels in Malaya and Scandinavian countries in geography broadcasts. He also gave lectures about his travels, and animal life, in public meeting halls. His son, Gilbert, remembers lectures with magic lantern slides lit by an oil lamp, as a visual aid.[34]

In the New South Wales Government Gazette for February 1922, there appeared this notice:

"His Excellency the Governor, with the advice of the Executive Council, has been pleased to approve of —

Mr David George Stead, formerly Manager, State Trawling Industry, being granted leave of absence without pay from 1st September, 1921, to 17th January, 1922, both days inclusive, and from the latter date his services being dispensed with, in terms of section 65 of the Public Service Act, 1902."

So when Stead returned to Sydney at the end of 1923, there was no job for him. His son, Gilbert, said "they just wanted to get him and they got him while he was away".[35] The decision was made in 1923 that the trawlers from State Trawling be sold and in 1924 this was effected by a non-Labor Government. Gilbert Stead added: "If you read the material that was written, particularly in the press, violently opposing the New South Wales fishing industry and demanding that it be closed down because it was uneconomic, that has to be seen in its true light in the circumstances where shortly afterwards, a private company called Red Funnel Trawlers set up and they functioned right through those two decades until the Second World War, and profitably too".[36] The most unfair outcome of Stead's dismissal was the refusal to grant him his superannuation.

In July and August 1924, David Stead was in Honolulu, as part of a New South Wales Government delegation attending the Pan-Pacific Food Preservation Conference. In 1925 he was employed again for a short time, having been appointed Commissioner of a Rabbit Inquiry whose objective was to investigate methods of controlling and exterminating the rabbits that were the scourge of the New South Wales countryside. He wrote a five-volume report on the rabbit menace, but from 1927 was out of work again.

Out of work, but tireless in his fight for nature conservation, in 1927 Stead was President of the Naturalists' Society of New South Wales, President of the Wild Life Preservation Society of Australia, and President of the Aquarium Society of New South Wales, as well as an active member of various other organizations such as the Australian Forest League, the Gould League, Geographical Society of New South Wales, and the Royal Zoological Society of New South Wales. But although he was certainly a socialist, he did not join the socialist organizations — he was never a member of the Communist Party for instance, or the IWW (International Workers of the World). Son Gilbert remembers that the first activity of any kind in which he joined his father was a march from Eddy Avenue down into the Domain on the 20th anniversary of the 1916 no-conscription vote "and it was a very big demonstration for a peace demonstration".[37]

Stead's greatest strength was a belief in himself. Christina Stead, although acknowledging that the character, Sam Pollit, in *The Man Who Loved Children*, was based on her father, wrote a less exaggerated portrait of him for the journal *Overland*, published in 1972, fifteen years after her father died. In this short story, "*A Waker and Dreamer*", she wrote:

> "David's appearance of whiteness, fairness and all that goes with it, dazzled himself. He believed in himself so strongly that, sure of his innocence, pure intentions, he felt he was a favoured son of Fate (which to him was progress, and therefore good), that he was Good, and he could not do anything but good. Those who opposed him, a simple reasoning, were evil. This was not his mother's work, but his own nature. He would sing certain songs, especially when something went wrong in the Department, or his work in the Naturalist Societies, some defeat, jibe, unkind joke; he would sing, 'Dare to be a Daniel, Dare to stand alone, Dare to have a purpose true, and Dare to make it known'."[38]

Stead's uncompromising political style was a product of the social radicalism of the late 19th century, and his political interests continued to reflect the major concerns of the left during his life, right up until his death in the 1950s. Christina said: "My father was a socialist, a state socialist . . . He didn't know anything about Fabius or Fabianism. He was just an instinctive socialist who believed in state socialism, as it suited him. But he was not a theoretical man at all".[39] Allen Strom said Stead was one of the first to identify conservation as a political issue. "I think he understood, back in the early part of the 1900s, that the only way in which you could ensure the management of wild life conservation, or nature conservation, was to ensure that you had legislation that provided for the correct administration. And he began to fight for this soon after the [Wild Life Preservation Society] was established in 1909, and the record shows that he was very nearly there when the First World War commenced, and then at the end of the war the Government brought down the Birds and Animals Protection Act — this was about 1918 — but of course it was much less than he had hoped — he had hoped of course that there would have been some sort of service to go with the administration."[40]

In 1929, David became an adviser and director of a new company; The Australian Whaling Company Limited. The prospectus, issued on July 25, stated on page 25:

> "We have the high authority of Mr David G. Stead, the well known International Fisheries Expert, and one of the greatest authorities on whales and the whaling industry, in stating that an equally good field — and one that has been proved — lies in the waters of Eastern Australia."

The *Sydney Morning Herald* reported on August 11, 1930, that "The Australian Whaling Company Ltd whose headquarters are in Sydney, hopes to be ready to commence fishing in May next year, and is establishing a factory on the north-west coast of Western Australia". The company also had plans to operate a shore station on the south coast of New South Wales, to purchase five whaling gunboats or chasers, and a vessel of about 12,000 to 14,000 tons oil capacity as a whaling factory ship. It was an ambitious project, and one which would have given David Stead a good measure of financial security, but by October 1932 the *Sydney Morning Herald* was reporting on the past year of inactivity forced upon whaling companies by the sudden collapse of the market at the end of 1931. In 1931 The Australian Whaling Company was not listed in the *Digest Year Book of Public Companies,* although the three New South Wales directors (of a total of five) were still included in the list of Directors in the back of the book. In the 1932 *Digest*, there is no mention of the Company or its Directors, and no further information on the venture could be found.

David Stead's daughter, Christina, had meanwhile grown up through her awkward teens, studied at Sydney Teachers' College to be a primary teacher (her father would not allow her to attend University), failed to find satisfaction in the classroom, and resolved to leave her unhappy situation at home. On March 28, 1928, having scrimped to save her fare, Christina was farewelled by her family as she boarded the *Oronsay,* a Royal Mail Steamer of the British Orient Line. She was travelling to London, third class, with a dream of having her talents as a writer recognized, her talents as a woman writer, and as we read in her novel *For Love Alone*, she was looking for love and acceptance. She did not know that she would never see her father or her step-mother again. She would return to Australia as a mature woman, a famous writer, but never really reconciled to the land of her birth.

"What was your ambition?" Louie asked Sam, "You know it, Looloo," he replied in a deep voice. "It is to be of those who spread the light, the children of light."[41]

Chapter 2 Notes

1. *Meridian*, Vol. 8, No. 2, October 1989, p. 101.
2. Hazel Rowley, p. 3.
3. Christina Stead, *A Waker and Dreamer*, p. 484, p. 482.
4. Hazel Rowley, p. 4.
5. D. G. Stead papers, ML MSS 5413. Notes by Gilbert Stead.
6. Hazel Rowley, p. 5.
7. Chris Williams, p. 6.
8. C. Stead, *A Waker and Dreamer*, p. 488.
9. Hazel Rowley, p. 13.
10. ibid.
11. C. Stead, *Ocean of Story*.
12. Chris Williams, p. 17.
13. C. Stead, TMWLC, p. 35.
14. Selected Letters.
15. Selected Letters.
16. TMWLC, p. 143.
17. Conversation between Hazel Rowley and Thistle Stead, January 15, 1987.
18. *Meridian*, ibid, p. 107–108.
19. C. Stead, *A Waker and Dreamer*, p. 490.

20. *Meridian*, ibid. p. 108.
21. *Sydney Morning Herald* 26.9.18 and *Daily Telegraph* 26.9.18.
22. Fourth Sectional Report of the Royal Commission to inquire into the Public Service of New South Wales upon the State Trawling Industry, New South Wales. Parliamentary Papers, 1st Session, 1920, Vol. 1, p. 48.
23. ibid. p. 49.
24. *Australian Wild Life*, Vol. 2, No. 2, December 1949, p. 34.
25. *Australian Wild Life*, Vol. 3, No. 3, March 1958, p. 5.
26. *Meridian*, ibid, p. 106.
27. C. Stead interview with Rodney Wetherell in *Australian Literary Studies*, Vol. 9, No. 4, October 1980, p. 446.
28. ML Stead Papers MSS 5715 Box 1/25. Letter dated December 12, 1921.
29. C. Stead, *A Waker and Dreamer*, p. 491.
30. Chris Williams, p. 26
31. TMWLC, p. 237.
32. TMWLC, p. 356.
33. *Meridian*, p. 103.
34. *Meridian*, p. 103.
35. ibid. p. 109.
36. ibid. p. 110.
37. ibid. p. 103.
38. ibid. p. 102.
39. R. Wetherell, *Australian Literary Studies*, Vol. 9, No. 4, p. 443, 436.
40. *Meridian*, p. 106.
41. TMWLC, p. 135.

CHAPTER 3

And Gladly Teach — 1925–1945

"Before we can lead our children along the path of truth, we must travel it ourselves . . . it is the rebels, those who cannot accept the dogma of the already-formed doctrines of their teachers, who will lead us out of this entanglement into which we have been led."

T. Y. Harris, 1941

MURWILLUMBAH Intermediate High School was Thistle's first country appointment, from February 2, 1925 at a salary of £218.8.6 per annum. "The only reason why it was Murwillumbah was that when we were filling in our forms saying where we would like to teach, I was the only one who could spell that name. I spent four delightful years there." There was another explanation — she wanted to be close to Mt Warning.

The first day was a disaster. She was already two weeks late in first term, due to illness, and had travelled up the coast by cargo boat to Byron Bay, dogged by sea-sickness. In what was to be a pattern throughout her life, she had failed to book any accommodation. From Byron Bay it was a train journey to Murwillumbah, where she arrived at 11 a.m., asking directions to the High School from the Stationmaster. It was one mile, on a blazing hot day; the lunch hour was spent in successfully securing some accommodation. The first afternoon at school was in a portable classroom with a class which had been given no lessons all morning; Miss Harris was given some Geography books. When silence was not forthcoming, the new teacher issued a warning, Detention, no one will be able to go until there is silence. "But there was no silence, only whispering, and at last they told me the gate would be shut at four o'clock and there were spikes on the top. I let them go."[1]

Thistle didn't lead an active social life in this country town. She was young and unsophisticated and didn't make the best contacts because, as she herself said, she didn't know how to. Exploring the countryside, especially from a botanical point of view, became her main interest, and Mount Warning was a particular attraction, and she climbed it several times. In May of that year an older teacher was appointed to the Infants Department, Daisie McCulloch, a woman of forty-five who became a great friend and support. But Thistle was extremely lonely, even though she loved the teaching, and had started a company of Girl Guides.

One Friday evening Thistle was alone in her lodgings, miserable because it was a long weekend and her fellow teachers had left town. A knock came to the door. It was David, who had travelled by train from Brisbane where he was engaged in work relevant to his position as Rabbit Menace Enquiry Commissioner for New South Wales. Earlier in the year he had been approached by the New South Wales Government to investigate methods of controlling and exterminating the rabbits that were the scourge of the farmers in New South Wales. That evening they went to the pictures. Saturday morning Thistle was on duty with the Guides, along with a "mad woman friend" who helped with the Brownies. That afternoon the three of them set out for Mount Warning, and the skies opened, giving them a damp and uncomfortable night. But Sunday morning was fine, and they made it to the top before returning to town. Thistle and David took a bath at their respective lodgings and at 7 p.m. David returned to meet Thistle for dinner. On Tuesday she was teaching in a room that had windows facing the entrance, and one small boy said, "Miss, there's a man outside". It was David, a last look before he took the train back to Brisbane, and then to Sydney. He was forty-eight, and she was twenty-three. The two kept in close touch by letter all through that first year, and although the letters have not survived, unpublished jottings written by Christina Stead, in preparation for *The Man Who Loved Children* refer to the affair between the father figure in her novel and his young friend, "my sweet Alison".[2] In these notes the father is referred to as Alan Fairchild, and his wife is Eva (or Girlie), rather than Sam and Henny as in the final version of the story.

Girlie intercepts a letter from Alison, and writes to her with strong advice to look elsewhere for a romantic attachment. Thistle herself told Hazel Rowley (Christina Stead's biographer) that Ada Stead wrote to her in the early days of her affair with her husband,[3] and David displayed an indiscretion, or perhaps naivety, in discussing his feelings for Thistle with his older children. Thus Christina's notes are more than likely to be based upon the actual events which took place, allowing of course for some degree of author's fancy. Rowley states that Christina's fictional notes give an account of the courtship that concurs with the broad outlines of Thistle's account much later.[4]

> "He met her, they went to a teashop and they talked earnestly of the subjects that interested them, and of their hopeless but wonderful sustaining love.
>
> "And in their scientific and humane societies she had met him again.
>
> "Now there appeared a new letter regularly addressed to Alan Fairchild. Her husband used to come down early and wait for the postman at the gate. But often the letter appeared on the mantelpiece. She took one of them from the clock, took it up to her room, slit it and read: My dearest Alan: I was too busy in the library with cataloguing to get away for the meeting. Please forgive me. Where can we meet? I hope you have not been so ill as last week; I worry about you and am always thinking about you. Just let us meet in the park and if I can hold your hands and pass my hand over your dear face you will feel better, I know. It is hard for me not to see you too. I feel the cramps of love. I need you as much as you need me. Why must lovers be parted? I can't understand it. But I am yours. May it some

day be true in every sense. It will be somehow. I am surprised at myself when I think how much you mean to me. And I thought I was so independent! With love, your 'sweet' Alison."

Girlie questioned her eldest son when he came home and was surprised to learn he knew about his father's friend. "They went on an expedition together, a naturalist expedition, all the people in the Country Walks Society did." — Girlie decided to write to Alison.

"My dear Miss Darling, I have opened your letter to my husband and I find that it is a love letter. Of course you know that he is a married man with a large family of grown and half-grown children and a wife living. I have no intention of giving up my home or of breaking it up; I do not know what you have been led to believe. Whatever it is, the truth is that my children come first with me, even if I had not myself to consider. I do not see what you can hope for. Are you going to waste many years on a hopeless affair, when you can never marry the man? . . . Whatever you feel now, you must stop writing to my husband and stop seeing him. This will be the best for us all as well as for yourself. But I must warn you that if you do not, I will take other measures, Yours truly, Eva Fairchild."

As previously arranged, Alison met Fairchild in the park and showed him Girlie's letter, at which he became furiously angry. Some time later he tells his three eldest children, one after the other, about camping out on a mountainside with his sweet Alison. "I know I can confide in you," he said to them. "I was walking out on the mountain under the trees with sweet Alison and I know that I was in love . . . And then later I spoke to her about how we felt and she told me she had loved me before . . . That is why she didn't marry. She had found out what real love was, and she was prepared to go through life without me, but she could not marry another man."

Thistle returned to Sydney in August for the spring vacation, and the romance became an affair. Her parents were about to separate and she was staying with friends, with a greater degree of freedom than usual. But her father somehow heard about his daughter's affair with a man old enough to be her father, and before the holidays were over, he fronted up to David and asked for his word that he would stay away from Thistle, at least for several years. Thistle too decided she should break with David, and sent back his letters and parcels. How long the break lasted is not clear, but before she left Murwillumbah at the end of 1928, Thistle had decided it was long enough. "I had a wonderful time feeling sorry for myself and I decided that was very stupid, and I arranged to meet him again and it was just the same as before, and that convinced me it was the real thing, and it was the real thing, and we had a very happy life."[5]

During the time when Thistle was not seeing David, she consoled herself by buying a second-hand car from a dealer in George Street, Sydney, a four cylinder Cadillac, "a beautiful car". She borrowed £40 from her old friend of University days, Hilda Butler. She told the dealer to send it to Murwillumbah, but he refused, saying she'd have to drive. "I can't drive," she said, and he replied, "I'll teach you, when are you going?" "Tomorrow." "Right, I'll teach you this afternoon." But Thistle thought better of this and asked him to ship it up. She later had lessons from a young solicitor friend in Murwillumbah who surely would have been horrified to learn what a hopeless driver she was all her active life.

Thistle gained an admirer sometime in 1926 during one of the school vacations. She was helping set up an exhibition for the Wild Life Preservation Society on the 5th floor of the Department of Education building in Bridge Street, Sydney, an area left vacant most of the time for events such as this. Nineteen year old David Darwin Stead had heard of Thistle, no doubt from his father David Stead, but this was his first actual encounter. "She was a very beautiful young woman, one of the most beautiful women I had ever met," said David Darwin seventy years later.[6] "She was qualified and accomplished and spoke his language," he added, "it wasn't just love". The young David was studying engineering at Sydney Technical College, and went on to become a metallurgist, a corrosion technologist, specialist in the corrosion of metals, but he inherited from his father a life-long interest in natural history and has lectured to many groups in his long life-time. "Only hundreds, not thousands like my father," he said. David Darwin said on his 89th birthday how his father would have enjoyed seeing his son become a F.I.S.H. (Fellow in the Institute of Sea Horses, an American body for people active in the study of marine corrosion of metals), but in 1956 when the award was made, David George had already taken to his bed in his last illness. David D. Stead also became an accomplished photographer, like his father.

Young David's opinion of Thistle was not shared by her sister, Nan, when they were growing up. Nan said Thistle lacked the feminine skills and graces, which Nan possessed, and so failed to attract young men as Nan did. Furthermore, Thistle had strong opinions, disapproved of idle chit-chat, and was intolerant of people with different ideas from her. The formalities which are part of our social fabric were not for her.

In June 1928 Thistle lost her friend, Daisie McCulloch, who received a promotion to Infants Mistress at Morgan Street School, Broken Hill, but Thistle followed her there at the beginning of 1929. She tells in one of her taped interviews of an extraordinary man in the Department of Education who was responsible for appointments; he would do what he thought best, either separating individuals or putting them together, and he often succeeded in placing friends together in country appointments, a great boon for lonely young teachers with families far away. In Broken Hill she was not as lonely as she had been in Murwillumbah because she actually had three teaching friends there, due to the kindly appointments officer, she believed. But her immaturity still surfaced.

> "Apparently I was very attractive to 17 and 18 year olds, and a few inspectors visited Broken Hill, and I understand they were quite disturbed about it. I had a motor car and these senior students took command of the car when I blew my hand up in the lab, much to their great delight."[7]

A much more satisfactory association was made with a reputable botanist in the district, Albert (Bert) Morris, with whom she had already been corresponding for some years. Bert Morris was a world authority, not only on the local desert plants, but on Australian plants in general, and he introduced Thistle to the western flora. In his home, Bert and his wife, Margaret, had amassed a herbarium of 7,000 different specimens of flora from all parts of the world, but the local desert flora formed the more substantial section of the collection. His greatest contribution was achieved when the Broken Hill Zinc Corporation sent for him to assist them in overcoming the problem of sand-drift which was menacing their new million-pound shaft and plant. This was in the mid 1930s, and during the four years before his death in 1939, he succeeded in combating the problem. His advice to the Corporation was simple. "Why not give Nature back a little of what was once hers," he told the management. "Plant trees — not just any kind, but her own native flora — and watch how, in her own miraculous way, she will give thrice-fold back to man." He was given a free hand by the Zinc Corporation management.

Albert Morris died at the age of 52, but he lived long enough to know his faith in Nature had been justified. He made possible the Broken Hill Zinc Corporation's flourishing 22 acres of fruit and ornamental trees known as the Albert Morris Park. Inspired by the success of Morris, other mining companies began fencing in areas, and planting trees.

In 1929 the young Thistle was at first doubtful about Bert's expertise because he didn't have any academic background, and she was still a bit snobbish about this. "He would take me out and point to a tree, perhaps a mile away in this clear air, and he would say 'That is a so-and-so,' and I said, 'Drive over to it.' After he'd done this a few times, I discovered that he really did know what he was talking about. Of course, I wasn't then sufficiently knowledgeable to realize that he had a very good knowledge of the geology and topography of the place and knew that there were only perhaps a dozen different kinds of plants that would grow in this locality, so it wasn't quite as difficult as

I thought. But he did have a wide knowledge and he took me to all sorts of places in the inland so that I became very familiar with that and very attached to it."[8]

It was Thistle's association with Bert Morris in Broken Hill which gave her an introduction to an appreciation of the significance of ecology. When she was in Murwillumbah, and even in all her excursions in the Sydney district, her main interest had been in the individual plants, and it was from this time in Broken Hill that her knowledge broadened to include a study of not only what was there, but why and how it was there.

One of Thistle's friends was Kath Elger, a Home Economics teacher in Broken Hill. She tells of how Uncle Bert (Morris) got them up one night at 1.30 a.m. to drive out into the desert to see a flower opening. It only flowered for two hours. He took them back to their lodgings in his Dodge at 3.30 a.m.[9]

Young Edith Noble (now Edith Winter) was a first-year student at Broken Hill in 1929, and Miss Harris was her English teacher. Young teachers in country high schools were often expected to teach out of their discipline, but Edith remembers Thistle as a good teacher. Edith's mother belonged to a Naturalists' Society which was conducted by Thistle for members of the community in Broken Hill, the main activity being excursions into the desert surroundings to study the plants. Edith accompanied her mother on one such excursion and still remembers her first encounter with Sturt's desert pea. The family transferred to Newcastle at the end of 1929, but Edith still has a farewell gift from Thistle — *Catriona,* by R. L. Stevenson.[10]

In early August of 1929, Thistle received a call from John Fairfax Limited, where her father worked, asking her to accompany him on a trip designed for rest and recreation, as he had been ill. Charles Thomas had made his way upward through the computing department to the assistant managership and finally the general managership after W. G. Conley's death in 1929. Gavin Souter in his history of the *Herald* newspaper described Harris as "a short roly-poly man who loved the pleasures of life, some of them perhaps too well. He was generous and even reckless with money, an enthusiastic gambler and drinker, and possessed of as quick a wit as anyone else born within the sound of Bow Bells".[11]

The trip, undertaken no doubt to alleviate Harris' symptoms of excessive drinking, was a three week cruise on a Dutch ship to the East Indies, including Bali. After a bad start, Father Harris didn't drink during the trip, Thistle reported, and for her it was a most interesting and enjoyable experience even though she went alone to some of the shore excursions. In September they were returning when one of the passengers, a suspected case of smallpox, was put off in Brisbane. The remaining passengers and crew were put into quarantine at North Head for three weeks. Thistle spent three enchanting weeks among the spring flowers in the midst of a very disgruntled crowd, most of them wealthy tourists dissatisfied with the delay and the moderate level of accommodation at the quarantine station. Kath Elger had driven Thistle's car to Sydney in the spring school vacation, but going to see Thistle at the quarantine station, could only talk to her through the gate. "All the people lined up for telephones, but I was out among the flowers . . . beautiful weather and a magnificent area for wildflowers. Father just sat around, he was contented."

It was back to Broken Hill then, but not for long, because in August 1930 Thistle was appointed to St George Girls' High School at Kogarah, a suburb of Sydney. In May 1931 Daisie McCulloch followed her to the city with an appointment to Ramsgate, and the two friends shared a room in a boarding house at 29 Villiers Street, Rockdale. Mrs Wright was the landlady of this establishment, a large house with four big bedrooms and a number of smaller ones. Living in the same house was a young country girl attending Sydney Girls' High School. Eva Ward (later Eve Levis) was the daughter of a teacher at a small school in Bonalbo, and rather than travelling to the nearest high school in Lismore, Eva was sent to stay with her mother's friend, Mrs Wright. Eva was expected to help with small tasks like cutting lunches, but all the older women teachers living there watched over her like foster-mothers.[12]

The dining room table was Eva's place for homework. One evening she was reading a light novel by Elinor Glyn called "*It*", a very sexy novel Eve Levis recalls. Thistle walked through the dining room, saw what Eva was reading, and reacted immediately. She was disgusted, took the book from the girl and said "Eva, you can't read this kind of tripe! I'll give you something better." She returned with an Oscar Wilde book, "*The Happy Prince*," and fortunately Eva enjoyed it. From then on Thistle supervised Eva's reading.

Back in Sydney Thistle was able to meet David Stead quite often. They would go out in her car, attend meetings, sometimes see a show, but more often just enjoy walking out together. "We often went over to Clifton Gardens."

A close relationship had continued with her father, the parents now having a trial separation, with Charlie living at the Metropole. Thistle would often visit him in his lodgings, park the car in the street, and sometimes stay overnight. She didn't notice the inconvenience of the low teacher's salary because she was subsidized by her father: he loved to buy her clothes. "Neither my mother nor my older sister would accept frocks chosen by him. I'd accept anything. He used to go into Farmer's or David Jones, get the woman in charge to line up all her staff, then he'd choose the one whose shape was closest to mine and she'd act as a model, and he'd send it off to me. He sent me some beautiful things. When he couldn't read my letters he sent me a typewriter, but he could never read the typing either. It never did occur to me to have lessons."[13] Eve Levis said that Thistle didn't care much about her appearance, and was careless in dress, not a good advertisement for David Jones and Charlie's generosity.

Thistle's older sister, Nan, could possibly have done with some support from their father. Nan had married Joe Clark in 1923, and a son Alan had been born in 1924, but the marriage broke up while Alan was still young. Poppa, as Thistle's father came to be called, did find a job for Nan sometime in the early thirties, a clerk's position in David Jones.

It may have been in the late twenties or early thirties that Poppa bought a block of land at Careel Bay on Pittwater, north of the city. It was a waterfront block of virgin bush, one acre, and on it he placed an old bus body in which they could camp. The bus was called "Fez Pliz". Christmas 1931 was a happy occasion, camped at Careel Bay, Thistle and a group of friends including Kath Elger and Daisie McCulloch. Poppa had bought the young people a launch, and Kath Elger remembers that they used to travel to Palm Beach each day by the water.

From about 1930 onwards, Thistle began her extensive travels in New South Wales and interstate, always with a single purpose — to extend her knowledge of the flora of Australia. With Kath Elger she drove to Melbourne one school holiday, just to look at the Botanic Gardens. Easter 1931 saw Kath and Thistle and other friends on a trip to Kiama, Nowra, Kangaroo Valley and Fitzroy Falls. In June of the same year a group of five or six women camped in the Burragorang Valley. Kath Elger was probably Thistle's first companion on her many trips to Kosciuszko — she camped there every January for decades. The love affair with the alpine area began in 1922 when Thistle went on a student excursion there one summer, staying at the old Kosciuszko Hotel (destroyed by fire in 1951). "We spent our time mainly on Pretty Plains. We didn't learn many new plant species — so strange to us all, including staff — but it stirred in me a deep desire to know more of this strange, fascinating place." [14]

Kath Elger picked most of the plants Thistle required, while Thistle photographed them and then put them in the plant press. She built up quite a large herbarium, and it was only later that she turned to photography as well. "I must say I was rather amazed at my own skill, after a good deal of lack of success and spoiling the film and so on." [15] One year while camping at Kosciuszko the weather was so cold that Kath told Thistle she would never travel with her again, and she didn't. "Thistle was tough!" [16]

In 1932–33 Thistle was obliged to ask for leave once again because of her father. He had been sent by Fairfax Limited to

London to buy some machinery and took ill while still overseas. Thistle was asked by the Company to go to London to bring her father back, and was granted leave from October 27, 1932 to April 19, 1933. Two of Harris' grandchildren affirm that it was family knowledge that Harris had taken his secretary on the trip to London. While in London, he was asked by the directors to resign on grounds of ill-health, and on his return, his marriage separation became final. His youngest daughter, Joan, then a teenager, was very upset by the turn of events; she had been attending school at Redlands, but became a boarder at Abbotsleigh School for Girls at Wahroonga. Thistle was very appreciative of the co-operation she received from her headmistress, Miss Lilian Geer, a lady who apparently was not well liked by most of the staff. Thistle said, "I had to get leave twice, the Fairfaxes fixed it".

Residence in the city of Sydney enabled Thistle to become actively involved in many conservationist organizations. She was on the Council of the Wild Life Preservation Society, of which David Stead was either President or Secretary, alternating these positions over a number of years; she joined the Australian Forest League and became interested in their young members, the Junior Tree Wardens, and compiled and edited the publication *The Junior Tree Warden* for its first release in May 1937, continuing as its editor for several years. In 1935 she became Honorary Secretary of the Illawarra-Bankstown Federation of Junior Tree Lovers following a move by Mr W. J. Campbell, Honorary Secretary of the Schools Branch of the Australian Forest League to encourage co-operative efforts between schools and between schools and other bodies. St George Girls' High School was the leading spirit for a tree planting in Stanley Street, Kogarah, the ground being prepared by Kogarah Municipal Council. On the afternoon of August 7, 1935 senior students took part in the planting of thirty-four white bottlebrush, and three red bottlebrush and two *Grevillea banksii* in a triangular bed at the top of the street.

In an issue of *The Junior Tree Warden* early in the 1940s, Thistle as Editor had written a piece entitled "What the Schools Branch Stands For".

> "The Schools Branch exists for the purpose of assisting teachers to train Australian children to a true understanding of their civic responsibilities by an appreciation of their own tree flora. In this way it is hoped that a new generation will arise, conscious of the beauty of nature around them and with a consequent desire to conserve what remains of the flora and to make a world lovelier to dwell in by the presence of more and still more trees. Any assistance or advice which teachers may require will be readily supplied by the Hon. Secretary, Thistle Y. Harris, Teachers' College, University Grounds, Newtown, NSW."

It is generally agreed that the first Arbor Day in New South Wales was held at Ryde Public School on Wednesday, July 16, 1890, when Lord and Lady Carrington planted trees in the school reserve. The idea for Arbor Day had originated in South Australia in 1889.

The *Sydney Morning Herald* for July 29, 1939, carried an article on Arbor Day:

> "At celebrations in the Kogarah district, a member of the official party, Mr F. T. Berman, organizer, paid a tribute to the late Miss Constance Le Plastrier and to Mr Edward Webster, the composers of some of the choruses sung by the children. He quoted from Miss Le Plastrier's verse about trees: 'For when we are dead and gone, they'll still be growing on, to show that we were good Australians'."

Constance Le Plastrier had died in 1938.

In the early thirties David Stead had acquired a position as General Editor of the Shakespeare Head Press Australian Nature Books. Constance Le Plastrier wrote "*The Story of our Plants: First Steps in Australian Botany*", published in 1933. It was intended to be a beginner's book in botany, but is quite scholarly and detailed and admirably refers to many Australian plants for examples of structure and function. In his editorial for this book, David Stead praised Le Plastrier as not only one of Australia's most experienced teachers of Botany to young people, but also one of its finest field naturalists. Stead himself also wrote two books for the series in 1933, *Giants and Pigmies of the Deep* and *The Tree Book*. Ten books were planned for the series but only five were ever published.

Thistle Harris had a lasting impact on the school and the district where she was teaching at St George Girls' High School. She initiated a local native tree campaign, and the students learned their botany in practical fashion. They learned to select trees suitable for the district, to plant them and care for them, and in a short time they had three sides of their playground shaded with fine specimens of blackbutt, spotted gum and cedar wattle. In 1933 she supervised the planting of one hundred lemon-scented gums along the length of Victoria Street in front of the school, and by 1939 the trees were already 20 feet high.

With the co-operation of the local Shire Council, senior students organized and implemented the project in both the school grounds and the nearby streets: they planned the lay-outs, selected the plants, prepared name plates for plants, arranged the social function for visitors, initiated the younger students in the art of tree planting and care, and arranged for every visiting adult to be accompanied by a student. After the planting the girls carried out an organized lantern-slide/talk in which several participated, each exhibiting the species they had planted. This was teaching with vision, a belief in hands-on experience which she was to espouse for years to come, a fine example from a young teacher just thirty years old.

And the lemon-scented gums in Victoria Street made history in more than one sense; when they threatened to interfere with the telephone wires, the local authorities decided to remove the wires rather than spoil the trees. Unfortunately these "Thistle trees" were removed many years later by a Principal of the school who was alarmed about dropping branches. Ruth Sutton, now a teacher at Wheeler Heights in Sydney, went to St George Girls'

High in 1948, ten years after Thistle had taken a position at Sydney Teachers' College, but her influence was still in the school. The teaching area she had established for teaching biology was still in use — wet and dry areas, sometimes called the Fernery, a natural area. The school was still very proud of her, and she was held up as an example of a great teacher with a lot of influence. Ruth Sutton is now an environmental educator; she has a mentor she has never met — it was Thistle Harris.[17]

But Gloria Watkins tells a different story. She was in Class 1A in 1931 at St George and Miss Harris was her teacher for Elementary Science. "The class gave her a bit of a rough time," Gloria said, "because we soon found out she didn't know much Physics and Chemistry, it was not her strength. One day the girls taught her what 'density' was." Gloria was later to be one of Thistle's colleagues at Sydney Teachers' College.[18]

However, the *Sydney Morning Herald* for July 1939, Women's Section, carried an enthusiastic article on Thistle's "Vision of a Garden City". The article reported that from the intensive activity at Kogarah, Miss Harris's influence had radiated out through the whole state. "And the model tree nursery which she established there has been copied many times since in other schools." Thistle was quoted as saying: "Sydney is not an easy place for street planting. But that is no cause for being discouraged. If we take the trouble to look for them there are many native trees which can adapt themselves to the city conditions. If I could have O'Connell Street to plant, for instance, I guarantee that I could make it beautiful with flourishing trees. A row of Sydney redgums would find themselves at home, or I could line the whole street with paper barks and train them to tidy city habits . . . Or give a windy street I would plant it with New England peppermints . . . And I shall have them yet," Miss Harris said with the light of enthusiasm in her eyes. The 1939 article concludes with this surprising story:

> "It must have been this same light which a Cockney flower-seller caught when he picked out Miss Harris walking along

wintry London streets. 'Australian eucalyptus lady' he called out as he darted across the road to her and pushed a bunch of gum-tips into her willing hands."[19]

Meanwhile changes had been taking place in her personal life. In 1931 Ada Stead, David's wife, had moved out of her Watson's Bay home and gone to live with her cousin Rachel Davis, at The Entrance. Her youngest child, Gilbert, was now a teenager and perhaps she thought her main responsibilities were over. Later in life, Christina Stead was not without sympathy for her stepmother. "She married the wrong man, that's all," she would say, "and that's extraordinarily difficult."[20] As a child, Gwen Walker-Smith was aware of the "feeling of tension in the air when David and Ada were together". Her aunt, she tells "had no idea how to manage on a limited budget. To get away from the children she would go to town and buy things the family could ill afford."[21]

David was now a free agent to pursue his romance with Thistle, who on her return from London, had taken up residence with her sister, Nan, and nephew, Alan, in Darling Point Road, Edgecliff. In 1934 the trio had moved to 129 Hopetoun Avenue, Vaucluse. Alan often became one of the party on Thistle's 5–6 week camping trips, usually with Kath Elger, Alan cramped in the back seat surrounded by specimens. "Nature was thrust upon me," Alan said, "but I did not become a nature lover. It must have been in my genes, because it was not my environment. We'd go all over New South Wales, to Kosciuszko, up the Namoi looking for platypus. We often camped on a private farm, never asking permission, and one day a farmer came down with a shot gun. And I was usually the one to be sent up to the farm with a billy can to get some milk."[22]

Alan Clark also recalled the visits David Stead made to the house in Vaucluse, usually on an evening when his mother was at work. David would go to Alan's bedroom to sing him lullabies. "He had a terrible voice, but he fancied he was a child expert, and he did like children, but it didn't work with me and I couldn't go to sleep."[23] Christina Stead once talked in interview about the family songsters: "I've always had a weak voice — unlike the rest of my family who are all great singers, I don't mean 'great' singers, but they sing like mad".[24] Charlie Harris must have become reconciled to his daughter's association with David Stead, because David would sometimes come camping with Alan, Thistle, and Poppa, down in the bush in Careel Bay.

David's home at Boongarre was still his castle, but the frenetic activity of previous years had given way to a different kind of domestic order. His daughter, Kate, kept house (although David did like to cook), son Kelvin was still around, but sixteen-year-old Gilbert had run away from home in 1933, stowing away on a ship. David Darwin had married his first wife, Percia, in 1932. David senior's job with the Australian Whaling Company had come to an end, and there was only his writing and numerous societies to keep him busy. *The Rabbit in Australia* was published in 1935. As was normal, money was in short supply.

On March 14, 1938, Thistle took up a position as lecturer in Biology at Sydney Teachers' College on a salary of £420.2.5 per annum. Ada Stead had been living with her son, Fred, and his wife, Mavis, in the Blue Mountains for some years. Christina Stead wrote to her brother Gilbert from New York in November 1937, "I hear Fred has become a speedster," and in August 1938 Fred was killed by a truck while riding his motor bike. He was just 27 years old, and left a wife and two children. David G. rose to the occasion and helped with arrangements for the bereaved family.

Thistle was welcomed to Sydney Teachers' College, together with another newcomer, Miss Anna Hogg, at a stylish afternoon tea in the College dining room, organized by Miss Skillen, Warden of Women Students. When Miss Skillen saw that the young women had come to the gathering without hats, she took hers off and put it under the table.[25] About this time, 41.5% of the staff were female. This differed greatly from the situation at the University of Sydney, where few women were employed as lecturers. Part of the explanation is that staff at College were appointed under Public Service Board regulations which, in this matter, applied equally to men and women. Since women were strongly represented in the teaching service it was only natural that they should also figure prominently amongst applicants for College positions.[26] Anna Hogg had graduated with first-class honours, and the University medal in Philosophy at Sydney University. She became a lecturer in Education, gained her PhD, and in 1948 took charge of her Department. Another newcomer in 1938 was M. L. G. Sheldon (Merv), who was appointed to lecture in agriculture and biology.[27]

Professor Mackie, the Principal of the College, placed the newly appointed Miss Harris in charge of the natural science section in the primary school curriculum. Thistle's own story tells of her first encounter with her Principal.

> "Now I'd never been in a primary school in my life except as a pupil, and I said to Professor Mackie, 'This is not really my strong point.' He countered — he was a little Scotchman — 'Nobody else wants to do it so you'll have to!' So I found myself a school in the district and said to the teacher, 'Would you let your children come over to the Teachers' College and I'll handle them over there and find out something about this?' and he said, Yes, and they came over for quite a while. The first time they were due to come over, I had the laboratory prepared with material. Professor Mackie's secretary rang me up at home one night and said, 'Oh, Professor Mackie has arranged for you to have your class in Room 102.' Now, Room 102 was a very special room that had a beautiful carpet on the floor, exquisite chairs and a special recording machine and all sorts of beautiful equipment, so I said, 'Oh no, I'll have my class up in the laboratory, thank you.' He parried, 'Oh, but you can't.' I said, 'Why not?' 'Professor Mackie has arranged for you to have it in 102,' and I said, 'I'm very sorry but I can't have it in 102, it's not convenient.' The secretary, horrified, said, 'But Professor Mackie *said* you must have it in 102,' and I said, 'I'm very sorry but I don't think Professor Mackie would like masses of petals and leaves trampled into his new carpet.

I'll have it in the laboratory, thank you,' which I did, and that was the beginning of my work at Teachers' College."[28]

Thistle was struck by the conservative atmosphere at Sydney Teachers' College. She wrote a comment, "Sydney Teachers' College 1940–1960" which was reproduced in the history of the College, and is here repeated in full, because it gives the reader a keen insight into the attitudes which were behind much of Thistle's aggressive behaviour at College, particularly in staff meetings. For many members of staff, she wrote, "the sun still never sets on the British dominions; a white Australia was accepted as an essential component of our existence: communists (if one dare breathe the word), must all be deported to Siberia with celerity. There was a hierarchical structure influenced by the kind of 'academic snobbery' which ranked in descending order: university lecturers, college lecturers, high school teachers and primary school teachers. Everything was dominated by a small group of 'privileged old and honoured members'; younger lecturers and newcomers figured scarcely at all. Students who attended University were similarly regarded as superior to their fellows. [But Thistle as a student herself held this view.] The whole structure solidified as the College grew older and attitudes became more deeply entrenched." Women, she complained, operated both within and outside this structure. They were promoted on an equal footing with men. On the other hand they rarely obtained positions of responsibility, such as headships of departments and senior administrative posts.

In early 1939 Thistle took a bold step and moved into Boongarre at Watson's Bay to share part of the house with David Stead. It must have been a financial comfort for David to have a steady income to help with his share of the expenses. Also in the house were David D. and Percia (upstairs), Kel and his wife, Ivy, just married, in one half downstairs, and Thistle and David in the other half downstairs. Two tiny rooms were set aside for Kate and Gilbert who were coming and going at this time. Thistle recalled that because she was living with a married man, the Watson's Bay locals tended to shun her, but she minded less than David did. David's sister, Florence

Walker-Smith, was very upset at the move, her daughter Gwen recalls, because she was so fond of her brother. Gwen said, "Thistle was a very pleasant woman who was fond of all Uncle Dave's children. She and Uncle Dave had a lot in common, but of course it was Aunt Tot [Ada] who was the mother of all his six children. It must have been very awkward for Thistle when she moved in with Uncle Dave."[29] Ivy Stead's opinion is that "nobody cared" except Thistle's father who, already very ill, came to the house and made a huge fuss.[30] Obviously his reconciliation did not extend to the idea of cohabitation.

Alan Clark recalls being aware of the fact that Thistle very much desired to have David Stead's child, even though they were not married. Thistle herself openly discussed in interview the desire they both had to have children. "I wanted to, very badly, we both did. He was quite sure any child we would have had would be much better than any human who had ever lived before — much more intelligent, much more pleasant — he was quite convinced, but I wasn't quite so sure."[31]

Thistle did not become pregnant, and she consulted a gynaecologist in Macquarie Street, Dr Norman Haire, who told her she was unable to have children. "He wouldn't do anything, and he wouldn't encourage me to do anything. The difficulty was I didn't know how we were going to manage it because we were pretty money shy and I couldn't have left work."[32] Thistle referred to Norman Haire as "a woman's doctor who was very outspoken". He was exactly the kind of person Thistle would consult, a man not accepted by the medical establishment although he had rooms in Harley Street, London, and had written several books on topics which included birth control, sex problems, sex practice, as well as an Encyclopaedia of sexual knowledge (1934). He was Superintendent of the Royal Hospital for Women, Sydney, 1916–17, served as a Captain in the First World War, held a number of hospital positions in the United Kingdom in charge of Birth Control Centres, and from 1936 to his death in 1952 was President of the British Sex Education Society.

But he was never given a written entry in the very conservative *Medical Journal of Australia*, probably because he was Jewish, homosexual, and wrote about sex.

In December 1939–January 1940, Thistle went on a five-week trip to the Centre of Australia, Alice Springs, travelling in her Chevrolet car via Adelaide. At the beginning of the trip she was thrilled to meet up with her young friend, Eva, now married to Ken Levis whom she had met at Teachers' College in Armidale in 1934–35. "It became Eve Levis — it flowed better than Eva Levis," she said. Monty Embury, who lived in Eastwood, organized trips for teachers and others interested in the environment. There were fifty people on this trip, in ten cars, A model Fords and Thistle's Chevrolet. They left the cars at Quorn and boarded the train — they called it the Dirty Ghan — travelling to Alice Springs and arriving thirty-six hours late. They stayed three weeks in Central Australia, then returned to Quorn and travelled back through Adelaide. "Thistle was with us most of the time, wonderful to us and our two male teacher friends. She could be very spiky but a stimulating friend," said Eve.[33]

Together David and Thistle threw their energies into their causes, writing letters, giving lectures, continuing research and acting in positions of responsibility in their various societies. In 1943 both of them wrote to the Education Department in Victoria to protest against the expulsion of some children from school because they refused to follow the ritual of honour to God, country, and the flag. Their protest claimed that the children could have little understanding of these "meaningless obeisances".[34]

A letter from David in February 1946 to Rev. H. W. Woodhouse of Chatswood runs, in part, "I have been drawn more and more to the side of the Communist organisations — although never a member of any — through one great outstanding thing: their unfailing support everywhere for human freedom from exploitation both by Imperialism and by Great Capital — which today are one and the same thing".[35]

However, David Stead's most vitriolic correspondence was kept for criticism of religion; he wrote many letters to men of religion, criticising the Bible, so-called Christian ethics, and the blight of religion on human progress. His papers include scores of clippings from newspapers and newsletters concerning religion, and one wonders whether he were not obsessed to an unnatural degree with opposing something with which he disagreed. A letter to the Rev. Cameron of the Bible Reading Fellowship of New South Wales, October 31, 1947, reads in part: "For the past 45 years I have made an intensive study of the religious beliefs of the peoples here and elsewhere, have studied the published statements of many, many Christians, have heard their confessions of faith in public and privately and have questioned quite a large number regarding the foundations and sources of their beliefs and what they considered to be their 'Christian attitudes'."[36]

Another letter was sent in late 1949 to the eminent science educator F. Sherwood Taylor, who had recently joined the Roman Catholic Church. A reply came back from Sherwood Taylor, from the Museum of the History of Science, Oxford, on February 22, 1950. "I really wonder whether there is any use in my replying [to your letter] because you are evidently so unsympathetic to religious belief that you would be unable to understand the motives which led to my conversion, and which maintain me in my present position . . . Frankly your attitude is one which, in my opinion, is out of date and based on a false conception of the nature of science."[37]

In February 1950 David wrote to the Rev. Dudley Hyde re Religion in Education: "Your attempted exaltation of religion and its place in education, in today's *Herald*, will, I feel sure, find no echo in the minds of most students of the science of education. I would say, after many years of study of the question that there is no place at all for the abominations of religion in educational processes, which ultimately stultify all freedom of thought in both child and adult; except in those fortunate cases in which the indoctrinated one has been able to escape from its horrid mental thrall . . . In fifty years fight against the evil influence of religion, experience has shown me that, although in many instances Press controllers have a complete contempt for those who preach religion to the masses, they willingly use them as allies to keep the people in moral and mental servitude — or as an aid to such . . . To many [teachers], as to myself, the Bible is anathema, with its absurdities, atrocities, immoralities, indecencies, and obscenities."[38]

While David was involved with his fight against ignorance in one quarter (as he saw it), Thistle had embarked on post-graduate studies by correspondence from Melbourne University. She took courses in education, gaining her Bachelor of Education after two years, 1943–44, and her Master of Education in 1945. "I don't think it had much value from an academic point of view or, what shall I say, a professional point of view, advancing my interests in the service, but it was very stimulating and I did make contact with some very attractive and interesting personalities, including a man at Melbourne Teachers' College, since dead, who was always known as Jumbo Wilson, with a great interest in natural science teaching in the schools and who taught me a very great deal. I chose for my thesis topic for this course the teaching of natural science in New South Wales schools."[39]

By 1941 Thistle had learned a lot about the place of natural science in the primary school, having reluctantly been turned in that direction by Professor Mackie. She expressed ideas ahead of her time in an article she wrote in the publication *New Horizons in Education* (September 1941), "Is the Pursuit of a Knowledge of Nature Worth While?" It is part of any appraisal of Thistle and her contribution that one should look at a few passages from this early article.

> "It is the mad-wag and the rebel who teach us the needs and the desires of humanity . . . The natural curiosity of children in things alive, things like themselves in that they move and grow, is, in many cases, effectively strangled in childhood, if not throttled at birth by unsympathetic parents (who themselves had their curiosity throttled by equally

> ignorant parents). It becomes replaced only too often by a complaisant satisfaction in the teacher's praise of a list of words neatly written and correctly spelt or a mathematical table committed to memory . . .
>
> "Now, there is no intellectual adventure of greater interest to the child than the exploration of his own widening environment. There is no single subject calculated to increase the child's knowledge and stimulate his observational and deductive powers more than this direct investigation — carefully guided — of nature around him. But — carefully guided? Aye, there's the rub. Because their own interest is strangled at birth . . . the teachers, schooled in books, are afraid of this immense subject. Quite rightly, they fear their own ignorance."

Thistle went on to suggest that it was through problem-solving and investigation that the child would become an independent thinker. But a capable teacher was needed to divert the child away from those aspects of his/her investigation which confused, towards that aspect which would give satisfaction. And the nature teacher was not to be wary of introducing the untouched fields of science into the primary school classroom. By careful guidance the teacher would lead his youthful students a step or two further along their road.

> "But how to use this in our present school system? And this, I fear, I cannot answer you. To train a child to seek the truth and to accept the truth only as it is revealed in the light of his own experience as part of a school training which prints, in the year 1941, that the aim of history teaching in the primary school is to teach the children "to see the slow but sure overthrow of the rule of Force by the rule of Right!" No — before we can lead our children along the path of truth, we must travel it ourselves.
>
> "The children of today are the teachers of tomorrow, and how can they, with their ideas handed out to them ready-made (however palatable the form), ever develop into leaders of thought? So we come back, like our education system, to

the beginning of our thesis, that it is the rebels, those who cannot accept the dogma of the already-formed doctrines of their teachers, who will lead us out of this entanglement into which we have been led.

"A study of nature, truly applied, could certainly lead us out. But it must be nature in the field and experiment in the laboratory which will solve our children's problems, not parables from nature or experiments from text books."

Thistle's conviction about the curiosity of small children was now based not only on first-hand experience with her visiting school children, but on a family situation at Boongarre. Kel and Ivy now had a daughter, Dorothy Ann, whom Thistle and David delighted in teaching and entertaining. Dotann, as she was called, would toddle into the back part of the house, early in the morning, to spend time in bed with David and Thistle. An undated letter, but certainly early in the forties, was sent by David to Thistle who was apparently away on a trip.[40]

Home
Monday Night

Beloved,

Bob was early this morning. I got your letter about 9. I am glad that you are having such a satisfactory time, with all the goings and comings — especially the short-comings. It will be very useful to you. I felt that would be so, when you decided to go.

The plants arrived (what a swag of them). I set them in four of the terracotta trays and placed them in the frame.

Sorry about the splinter. Hope it is better now. A miracle happened: twice in four days Dotann has come in while I have been having my early morning tea and has snuggled up against me in bed, just like she used to with you . . .

Dearest, I am counting the days. I love You Terribly, Beloved For-Always-and-Always-and-Always The Love That Passeth All Understanding

The Same Yesterday To-day and Forever

David

Thistle had made her decision about her allegiance to David Stead, one from which she never wavered, or ever would. The 1940s and the years that followed were to see her involved in writing and lecturing on behalf of Australia's wildlife, but the most important contribution she was to make in the world of publications was her first, the 1938 book, *Wild Flowers of Australia.*

Chapter 3 Notes

1. Details of the time at Murwillumbah came from the taped interview with T. Y. Stead for the Bicentennial Oral History Project, Catherine Johnson interviewing. ML Stead Papers MSS 5715 Box 2(25).
2. Christina Stead Papers, NLA, Box 1, Folder 3.
3. Hazel Rowley, p. 58 and 571.
4. ibid. p. 572.
5. Taped interview, MSS 5715 Box 2(25).
6. Interview with David D. Stead, 4.10.96
7. ibid.
8. De Berg Tapes, January 1972, NLA, Tape 574.
9. Interview with K. Elger, 7.2.96.
10. E. Winter, pers. comm. 12.6.97.
11. Gavin Souter, *Company of Heralds*, p. 147.
12. Interview with Eve Levis on 20.2.97.
13. ML MSS 5715 Box 2(25), Taped interview.
14. Letter to Neville Gare, 15.4.85.
15. De Berg Tapes, p. 6.

16. Interview with K. Elger, 7.2.96.
17. Interview with R. Sutton, 16.1.97.
18. Telephone conversation with G. Watkins, 19.2.97.
19. S.M.H. July 24, 1939, Women's Section.
20. C. Stead interview with Rodney Wetherell, broadcast February 24, 1980. Published in *Australian Literary Studies*, Vol. 9, No. 4, October 1980, pp. 431–8.
21. Hazel Rowley interview with G. Walker-Smith, Rowley's book, p. 27.
22. Interview with A. Clark on 23.1.96.
23. Telephone conversation with A. Clark on 7.2.96.
24. C. Stead Interview with Rodney Wetherell in *Australian Literary Studies*, Vol. 9 No. 4, October 1980, p. 432.
25. Telephone conversation with Dr Anna Hogg, 24.8.96.
26. *Sydney Teachers' College — A History 1906–1981*, p. 74.
27. ibid.
28. De Berg Tapes, p. 6.
29. Interview with G. Walker-Smith, 2.11.96.
30. Letter from D. Basili, 3.12.96.
31. Tape in Box 2(25) ML MSS 5715.
32. ibid.
33. Interview with E. Levis 20.2.97.
34. Stead Papers ML MSS 5715 Box 1(25).
35. ibid.
36. ibid.
37. ibid.
38. ibid.
39. De Berg Tapes, p. 7.
40. Letter supplied by D. Basili.

CHAPTER 4

Early Publications — 1938–1954

"I was impressed by her wildflower book — an outstanding achievement — a watershed — she encouraged people to understand how plants were classified, in a simple way. This was before I met her in person."

Allen Strom

THISTLE'S first book, *Wild Flowers of Australia* was produced in 1938 and revised in 1979 for its eighth edition, running to reprints in 1980, 1986 and 1989. But it was actually David Stead who conceived the idea for this publication, and it was he who was instrumental in persuading both the artist, Adam Forster, and Thistle, the author, to take out a contract with Angus and Robertson. It was back in the mid-1920s when David Stead persuaded George Robertson, managing director of Angus and Robertson and a man tremendously interested in Australiana, to commission Adam Forster to do the paintings which later appeared in *Wild Flowers of Australia.*[1] Adam Forster had been painting the Australian flora as a hobby since his emigration to Australia from South Africa in the late 1890s. David and Thistle knew him well during the 1920s when he was a member of the Naturalists' Society. In 1916 Adam had embarked on an ambitious project, his objective to complete 1,000 paintings of Australian wildflowers, but he died in 1928 when 918 had been completed. These magnificent watercolours, on paper 39 cm × 28 cm, are now in the care of the National Library, Canberra, placed there in 1949 by his daughter Mrs Elsa McGregor. Dora Forster, another daughter, used to accompany Adam on his wildflower trips in the years 1916 to 1927, collecting sites ranging from Ashfield to Dee Why, to Goulburn, Shoalhaven, the Botanic Gardens. Some of his paintings were done in the field, but most of them were painted at his home in Ashfield, or Strathfield, or later, Dulwich Hill.[2]

Adam had an interesting background. He was born Karl Ludwig August Wiarda on April 5, 1848 in Emden, Germany, where his father was a lawyer, judge and senator of Emden. The sixth child in a family of nine, Karl Wiarda had an interesting and varied life. He studied medicine in Emden before going to the Franco-Prussian war as a young man. He was a Prussian Guards Officer and was awarded the Iron Cross, a treasured memento held today by his grandson, Dick Forster. Dick also has a signet ring with the Wiarda family crest of a swan with an iron spiked ring around its neck.

Karl travelled to South Africa after the war, and ended up in the small town of Port Elizabeth in the Cape Province. There he met Mary Emma Smith, the daughter of a prominent citizen, William Smith, who was the first Mayor of Port Elizabeth. Records from Port Elizabeth show Karl as a book-keeper in 1882, and a merchant in 1885. Karl and Mary had three children, Frank (Florens), Elsa and Dora. He travelled back to Germany on several occasions while living in South Africa, but in the 1890s went to Australia without his family on the famous clipper, the *Cutty Sark,* his family following some ten years later. On the ship he changed his name, as the Boer War was in progress and he thought Wiarda would cause problems. He chose the name "Forster" and as he was to be the first Forster in Australia, he chose "Adam". Family memories about the choice of Forster include a story that it was the name of an accountant who died on the same ship, or, being interested in natural history, he chose the name in memory of the Forsters, father and son, who replaced Joseph Banks and Daniel Solander at the last moment as botanists on Cook's second voyage, 1772–1775.

Adam was described as a very intelligent, soft, gentle man, who spoke about six languages, and had the ability to look at a six figure column of figures and write down the answer. In

Australia he was Registrar of the Pharmacy Board and Secretary of the Pharmaceutical Society.[3]

A set of miniature paintings by Adam Forster was commissioned especially for the book envisaged by David Stead. Adam used to take them into Angus and Robertson in batches, and on one such trip in the tram, when he was in his eightieth year, he collapsed and died. It was April 11, 1928. His granddaughter, Betty Coldham, reports that he swung up on the tram and later died of a heart attack.[4] These beautiful little paintings are now in Angus and Robertson archives, under the care of Harper Collins Publishers at their premises in Pymble, New South Wales.

The plates are set out in sets of four, pasted on board of approximately A4 size, the paintings obviously having been cut out of a larger page of artist's paper. The plates are now somewhat faded, especially the reds and blues and greens, but they have been committed to film. In the earlier editions of the book, the colour reproduction was not always faithful to the original painting, but the 8th edition appears to be a more accurate rendering of the art work. For example, *Grevillea punicea*, in the 1947 edition, has too much blue in it, but Forster's original in Plate 16 has more of the red colour of the actual species, so the colour had been changed in the reproduction.[5] Adam painted two-hundred and forty-eight miniatures for the book.

About the time Adam Forster died, George Robertson also died. The blocks had been made for the book, and Edwin Cheel, chief botanist of New South Wales, had been commissioned to write the text. On the death of the two key people, the blocks and the handwritten text were put in the strongroom, where they stayed until 1937 or early 1938, when George Ferguson, who was George Robertson's grandson, found them. Young George had his grandfather's love of publications and of Australiana in general. He recovered the material, examined it, and returned it to David Stead, who suggested that Thistle could be consulted. Thistle was not aware that the consultation could lead to authorship; she was given the Edwin Cheel script by W. G. Cousins, who was the managing director of Angus and Robertson, and asked to give her opinion of it. In Thistle's opinion it was a technical, academic piece of work which would not have appealed to the general public. She recalls, " 'No, this isn't what is needed,' so Cousins said to me, and he mumbled when he talked, 'Will you do this for us?' and I said rather hesitatingly that I would. He had in mind quite a small publication, and I had in mind nothing at the moment, and he said to me, 'Would you be satisfied with £15 to do this job?' He mumbled so badly I thought he said £50, and anyway I didn't know anything about money values, so I said yes, and off I went. When I started to do the script I decided that it ought to be a book for students as well as for the general public and I prepared a second part, which is a key, and which involved a great deal of work."[6] Edwin Cheel did not die until 1951 and one wonders how he felt about the rejection of his work.

Thistle continued her story in the 1972 de Berg interview: "I said to the publisher, 'These should be arranged botanically.' 'Oh yes,' they told me, 'that will be quite alright.' So I arranged them all botanically and all the work was done. I was holidaying at the moment, somewhere in the country, and I had an SOS from A. & R. saying, 'We can't do it in this way because the paintings are made on different background colours, and we must put all the whites together, and all the greys together, so will you please reorganize all the work, and we want it by next week.' Added to this difficulty, I had a very bad cold, I was sniffling and watering at the eyes, and I can remember distinctly sitting down and working for very long periods, trying to get this done to time, because I was pretty innocent and I didn't have the sense to say, 'No, it can't be done in the time, you'll have to wait another week.' That was the beginning of *Wild Flowers of Australia*." A reviewer writing in *New Horizons in Education* (December 1944) complained that the arrangement of the coloured illustrations "has been dictated by the exigencies of the colour-printing process used, and the orderly arrangement of the second part of the book receives some mortal wounds in the first, where the most miscellaneous sets of four pictures are grouped on each plate". This reviewer also complained that the colours were reproduced with varying degrees of accuracy, and that many of the illustrations lacked distinction. However, "it will be very useful, especially for the vegetation of the eastern States".

When Thistle first examined the miniatures painted by Adam Forster for the book, she was critical of the selection, as not being a representative sample of the Australian flora, so they had a look at Adam's larger paintings, then still in the hands of Adam's daughter, Mrs Elsa McGregor. They decided on several additions, and Neville Cayley was engaged to reproduce some miniatures from Adam's larger paintings. Cayley was a well-known artist of natural history objects, especially of birds. The first painting he did was obviously a Neville Cayley painting, Thistle said, and she told him, "No, we don't want something that has a Neville Cayley character, we want something that will not be distinguishable from the original Adam Forsters". Thistle was eventually pleased with the Neville Cayley paintings and said she defied anyone to pick out the paintings that were done by Neville Cayley from the paintings that were done by Adam Forster.[7] It is possible to suggest which ones they could be; for example, one can identify the fragment taken out of the larger *Grevillea oleoides*, and the larger painting of *Acacia suaveolens* is identical to the one in the book. No reference is made to Cayley's contribution in the Foreword to the book.

The 1938 edition of the book was a small one, 22 cm × 14 cm. When George Ferguson came back from an overseas trip during the war years, he brought out a second edition, a larger book, almost quarto size, 25 cm × 19 cm. This 1947 edition of course had larger illustrations and the reproduction of the coloured plates was of better quality and clarity. Thistle herself said in 1972, "The first edition was very poor, the printing of the blocks was shocking and they blamed the war years for this. The whole thing was a bit of a mess. With the larger edition George Ferguson put me on a royalty basis".[8] In both the 1938 and 1947 editions Thistle acknowledged the help of Mr R. H. Anderson, Government Botanist in New South Wales, and Mr H. M. R.

Rupp for distribution figures for the orchids in the 1947 edition — the details are different in the two editions. No doubt she was also indebted to Rupp's advice re the correct spelling of the orchid *Diuris* in this later edition. In January 1943 Mr W. G. Cousins, the managing director at Angus and Robertson, had written to Thistle to say, "We do not know when we will be able to reprint *Wild Flowers of Australia* . . . We are not obliged to pay you any more money on account of this work, but will pay you twenty guineas for going through the current edition and fixing up any mistakes that have been made".[9]

In October 1944, Thistle wrote to Mr Cousins to say that with the demand for coloured illustrations of Australian plants and flowers, perhaps Angus and Robertson would consider the purchase of Adam Forster's collection of large paintings with a view to further publication. Cousins replied that the publishers were not particularly anxious to purchase, because if they did, they could not do anything with them until a few years after the war. However, in July 1945 George Ferguson wrote to Mrs McGregor, Adam Forster's daughter, asking to see the paintings because they were considering the possibility of publishing a more extensive book than the 1938 one after the war. However, nothing seems to have been done about this proposal.

Thistle's 1938 book on Australian wildflowers, and its later editions, made a great impression on enthusiasts of Australian natural history, but it was not the first book in this genre. In 1913 Florence Sulman brought out her book, *A Popular Guide to the Wild Flowers of New South Wales* with 51 full-page plates by Eirene Mort and Dorothy M. Watkins, Angus and Robertson. Volume I was released in 1913 and Volume II in 1914, this second volume having 72 full-page plates by Eirene Mort. Mr Edwin Cheel was mentioned in the Preface for his valuable help in field work. The notes on all the plants are detailed and scientific yet easy to follow, with notes on pollination and other adaptations. Interesting information on the historical background to the plant names was included, and a useful addition to the glossary was a series of illustrations for many of the terms.

While these two volumes by Florence Sulman cover many more species than the 1938 book by Thistle, the study is restricted to New South Wales, and the plates are black and white drawings, even though executed with skill and accuracy. The valuable contribution to Thistle's volume lay in the coloured paintings of Adam Forster, and its floral examples were drawn from all over Australia, giving it wider appeal.

Meanwhile, in her professional work at Sydney Teachers' College, Thistle was becoming heavily involved in the New Education Fellowship, a very progressive and influential body, and she became the Honorary Editor of its publication *New Horizons in Education*, in 1940. The New Education Fellowship was founded in 1915 as a rallying point for people of all countries who felt that a radical reform of education, based on a proper understanding of childhood and of the unity in diversity of mankind, was essential if ever world peace was to be assured. Since 1921, when its first Conference was held, it claimed to be the one educational organization which had organized sections and groups in all the major countries of the world and continued to follow a consistent educational and international policy. Before the war the Fellowship had sections and groups in 51 different countries, and 21 magazines in 17 languages were associated with its work. Its direction was in the hands of an Executive Board consisting of twenty-one persons from twelve different countries.

Thistle remained the Editor of the New South Wales publication of *New Horizons in Education* until early 1944. In May 1944 the six State sections of the N.E.F. in Australia were welded together, and the July 1944 edition of *New Horizons in Education* was put together by a South Australian Editorial Committee. In the 1960s the N.E.F. became the World Education Fellowship, and in Autumn 1967 *New Horizons* became *New Horizons in Education, The Journal of the World Education Fellowship in Australia.*

The New South Wales section of the N.E.F. was suspected of being leftist but it was not officially or openly so. It was most active in the 1940s and was considered by some observers as being used by a number of people as a platform to promote their own ideas. Through its publication, *New Horizons in Education*, Thistle certainly made clear, in editorials and articles, her views on religion in education, sex education, the primary school social studies curriculum, and the importance of learning from first-hand experience, as in Nature Study.

In the April 1943 issue of *New Horizons*, Thistle discussed some aspects of Social Studies in New South Wales Primary Schools in an article, "Are we Becoming Civilized?" The Social studies syllabus embraced History, Civics and Morals, and Geography, the aim of the syllabus (printed in 1941) stated to be "to produce men and women able to live in a land of freedom, justice and human happiness". Thistle's main objection to the History section is summed up in her words "Are we really teaching history when we tell them pleasant little stories about noble men and leave them to figure out why such noble creatures have left the world in such a mess? . . . We all know as historical fact that all the empires of the past and the present, and all the great nations, have been built up through the exploitation of the subject peoples, not excluding the mass of the population of the successful nation . . . It seems fairly clear that with such an unreal attitude the details of the social studies will hardly be such as will give the child that clear picture of his environment, past and present, essential to his development as a good citizen."[10] Although a colleague at the Teachers' College once called Thistle "a person of no admitted competence in the field of social studies," she nevertheless felt committed to extend her principles in that direction. A comment she made elsewhere supports her comments above, "It is essential pupils should know as much about the scoundrels who have caused pages to be written in history as about its heroes and idealists. A programme of history written by scholars nurtured in the lap of capitalism must inevitably laud those men who made the world 'safe for capitalism' and forget those who struggled for a free people."[11]

In the same article on Social Studies from *New Horizons* above, some strong criticism is levelled at the Scripture component of the Social Studies Syllabus. "Scripture deals with the Christian Scriptures, and there is nothing to indicate whether this should be regarded as history or as civics. As history, in the proper sense, it has no place, since we have little reason for believing that Old Testament stories or New Testament stories (with their continual contradictions) have sound historical foundations . . . I would not be very hopeful of the good citizenship of a group fed seriously on the mischievous tales from the Old Testament and doubtful of the ethical development of those who took the whole of the New Testament teaching as a basis for their behaviour, with its promises of rewards for services rendered and its punishments for lack of faith . . . It seems obvious that Scripture stories have no place either as history or civics. If anywhere, they might find their place in the folklore section of the English programme, where carefully selected stories from the Scriptures might be used. It would be a much more honest policy to substitute for this section a study of comparative religion . . ." [12]

In 1943 Thistle was editor of a slim volume, *Religion in Education*, which contained five addresses delivered before the New Education Fellowship (New South Wales). Two addresses were delivered at the first meeting. The Rev. Eris O'Brien pointed out that the objectives of the New Education Fellowship require "a greater understanding of the spiritual needs of all citizens," as he had read on the inside over of the magazine *New Horizons in Education*. "Had I read only that page of the cover of your magazine I should have decided that religion or no religion in education did not arise among you as a debatable issue," O'Brien said.[13] The Very Rev. J. C. Thompson said, "Now, of these two things — Religion and Education — Religion is the greater, for Religion is the end, and Education is the means, or one of the means thereto".[14] John Anderson, Professor of Philosophy in the University of Sydney, said at the second meeting, "the more religious instruction there is in any 'educational system,' the less is it truly educational".[15] The Rt. Rev. E. H. Burgmann, Bishop of Goulburn, had been the speaker at the third meeting, and on that occasion the hall was packed to overflowing. Dr C. R. McRae, Principal of Sydney Teachers' College, said at the final meeting, "The simple truth is that I do not know the answer . . . It is only too true that current education lacks an essential central core, a central unifying principle. Among primitive peoples, that need of a philosophy of life, of a point of view on the universe, is met by initiating the young into the religious beliefs of the tribe. The adolescent is given a religious explanation of the universe. Our questions is — what to do for the civilized youngster? . . . Finally, I would like to repeat that, whatever happens to religious education, our schools can and must give better education in social living."[16]

In the Preface to *Religion in Education*, Thistle as Editor does not venture into the debate, but points out that the Fellowship looks with favour rather than disfavour on controversial topics, "realising that the fullest use of intellectual resources can be made only if those forces are unleashed on a listening world".[17]

Thistle wrote a review of a book on "Sex Education for Girls from Eight to Twelve" in *New Horizons*, October 1945. "It must be particularly disappointing to teachers," she wrote, "to pick up another nauseating little volume to place with many of its predecessors in the dustbin". Her main criticism of the book was what she called "the confusion of biological information with the writer's idea of God and nonsense about such things as the sort of woman 'God wishes you to be'. Why sex biology is always selected for this kind of unhealthy association is a puzzle," she wrote. She went on to say, "One can write about respiration or digestion calmly and unemotionally and with no thought of what God thinks about it or wants us to do about it, but as soon as sex anatomy or physiology is mooted it must be carefully wrapped in some confusing religious apparel . . . I have no hesitation in saying that I think the book defeats its own ends. The child of eight to twelve years is forthright and impatient and does not wish to be taken on a long journey to get a few simple facts. She wants her information 'straight'."[18]

In sharp contrast to the review above was one written by Thistle about a book on birth control by Dr Norman Haire, the gynaecologist she had consulted about her own infertility. The book was entitled, *Birth Control Methods. Contraception, Abortion, Sterilisation* by Norman Haire, 1945. Thistle wrote, "For teachers the book has a special interest since it gives so clearly the point of view of a doctor who has had wide experience and has given much thought to a matter of considerable public concern. Those who know Dr Haire know him for his unequivocal conclusions based on his experience and his interpretation of human wants. Teachers' College students, I am told by a medical officer who lectures to them, are interested in contraceptives and would like to be told about them. Reproduction, in all its forms, still remains one of the most fascinating stories nature has to offer us. Contraception is one aspect of its study and is bound to be interesting to the study of biology of human affairs, and of other related subjects. Teachers, therefore, should know something about it and they can probably learn all they would need to know from this book."[19] This review was written in 1947, many years before it was acceptable to include sex education in the schools or tertiary institutions.

From early days in her teaching career Thistle had been a member of the Teachers' Federation. She was a staunch unionist, yet spoke openly against the policy of compulsory unionism. A letter she wrote on this issue was published in *Education* in 1942:

> "Democratic communities aim at organising their society for 'the greatest good for the greatest number'. For that purpose unions of groups of people working under similar conditions are formed, and these unions are concerned with watching the interests of their members and the particular trade or profession which they represent. This is good, and it is my firm belief that much more can be done by such an amalgamation of forces than can be done by individual effort. I know, for example, that the Teachers' Federation has done much both for educational progress and in the professional interests of its members. Therefore I am a unionist.

> "But because I believe in unionism I do not believe either in compulsory unionism or preference to unionists . . . The Teachers' Federation, you will say, should watch the interests of its members. The Teachers' Federation, I say, should watch the interests of the most highly-qualified or most-promising members, so that the profession is best served. No teacher worthy of the name would fight for the replacement of an extremely efficient headmaster by a less-efficient man merely on the ground of Federation membership . . . I believe that if the Federation will put forward a bold and forthright policy, it will need to use neither compulsory unionism nor preference to unionists."

Thistle closed this long letter on a note which betrays her own belief in the rights of the individual, for it was always as a self-sufficient individual that she lived and worked, even within the boundaries of an organization. "It is true that democracy implies the right of the majority to compel the minority. But the spirit of democracy is to find a method of living which will satisfy all its members. The higher the degree of intelligence in a group, the less compulsion should be necessary to ensure intelligent co-operation, and if we haven't it among our teachers, heaven help our democracy!"[20]

During the 1940s Thistle continued to involve herself with the younger members of her family, the children of her sisters, and the grandchildren of David Stead. Her younger sister Joan had married Peter Hunt on April 30, 1938, and a daughter Carol was born in 1940. When a younger sister Wendy was born in 1945, it was Thistle who looked after five-year-old Carol. Dorothy, or Dot-ann, the daughter of Ivy and Kel Stead was also born in 1940, and being in the same house made a close relationship not only possible, but desirable. Thistle was fond of Joan's girls too and often took them on outings, even when they were small. During their teenage years, Thistle would take Carol and Dorothy to the ballet every year, with a visit to Boongarre afterwards for Carol, and an overnight stay. "The house was always full of people coming and going," Carol said, "I never knew who they were but thought they must be Stead family".[21]

Carol's parents, Joan and Peter Hunt (Thistle's sister and brother-in-law) did not have much regard for David Stead because he did not work, and they believed he was responsible for introducing Thistle to his "strange socialist ideas."

Among Thistle's papers made available by her nephew, Alan Clark (son of Thistle's sister Nan), were a few typed chapters of a manuscript obviously intended for a book which was never published. One chapter, "Names for Things", had drawn heavily on her association at home with the young Stead children. She claimed that technical terms and names afford little difficulty, especially if one becomes accustomed to them from an early age. Dot-ann had been brought up on such names and showed neither fear nor amusement of them. From about the age of eight months it was a daily morning ritual for her to be carried out to one particular bush of *Melaleuca hypericifolia.* Thistle said she could give no other name for it because its only common name could be "Red-flowering Tea-Tree", which was clumsy and non-descriptive since several genera of Australian plants — *Melaleuca, Leptospermum, Kunzea* — are all entitled to the name Tea-Tree and many are red-flowering. At the bush, always spoken of as *Melaleuca* Dot-ann was given a small spray which, when crushed in the hand, gave a refreshing odour of essential oil. One of the first words she ever attempted to say, at the age of about 10 months, was *Melaleuca.* At first it was impossible to interpret it at all, a little later it became "Melamoocoo".

Another instance of the ease with which words can be accepted and used, Thistle claimed, was shown in Dot-ann's acceptance of technical names which she met in her constant perusal of *Natural History* (publication of the American Museum of Natural History). One afternoon when she was a little over three years old, it was not surprising to hear her say, "Will you read me about Brontosaurus — not Tyrannosaurus — but Brontosaurus". Thistle remarked that this looks pedantic in writing, but actually it was not. These were the names of things she had found of interest and they had been repeated a sufficient number of times to make them familiar. Dot-ann's brothers, John, born in 1943, and Robert, born in 1945, were also young students for the patient Thistle, and when old enough, all three were taken to the seashore by Thistle and David (or Bardie as the grandchildren called him).

The young Stead children always called Thistle "Missy". John Stead said he doesn't know where this originated but believes it may have come from David who had a nickname for everyone. When John's eldest daughter, Joanne, was small (she was born in 1976), she couldn't get her tongue around "Missy" and in the way of children started using "Mimi". John believes Thistle herself may have preferred "Mimi" because in all her correspondence with John's family from the late seventies she signed herself "Mimi". Thistle was to show the same patience with John's daughters (Joanne, and Alexia born in 1977) as she had with the young ones who grew up at Boongarre in the forties and fifties. John's girls "helped" Thistle for hours in her glasshouse, and she never objected when one of them took her a book to be read. John Stead remembers his childhood at Boongarre as a happy time but claims that his grandfather David was a greater influence than Thistle. He adds that this of course could have been deliberate, with Thistle not wanting to pursue a too active role as she was the new step-mother/grandmother.[22]

Thistle had a long and stormy association with Allen Strom, whom she met in the early 1940s. Allen was teaching at Enmore Activity School when he decided to take about 200 of these youngsters to the Broken Bay National Fitness Camp. Enmore Activity School was a special school for boys with learning problems, boys who had been educationally disadvantaged and were usually behind in reading and mathematics. At Broken Bay the children slept in tents and spent a large segment of their time in studying natural history (having been badly sunburnt on the first day). Later Allen produced a number of booklets relevant to the natural history studies at Broken Bay, and when Gordon Young, Director of Physical Education, saw them he arranged to have them printed.[23] Allen Strom had also set up a club whose main objective was to involve the Enmore Activity children in bush activities at weekends. This was the Caloola Club (meaning "to climb high") and the kind of people who joined in its activities were friends and relatives of the pupils, people Allen knew from his days at Sydney Technical College, and then later, when he lectured at Balmain Teachers' College (1947 onwards), students from Balmain. Thistle also belonged to the Caloola Club and often brought along students from Sydney Teachers' College. Allan Fox, who was later to be on the staff of the Broken Bay National Fitness School Camp, was a student at Balmain in 1948–49 and he recalls that he went out with the Caloola Club every weekend during those two years.[24]

The National Fitness Council used to meet in Sydney in the early 1940s and Allen Strom was frequently in the office. One day David D. Stead (Thistle's stepson), whom Allen knew from the Sydney Bushwalkers, came into the office and was shown the booklets Allen had produced for studies at Broken Bay. David D. showed the books to Thistle, who was immediately interested, and she invited Allen to visit her at Sydney Teachers' College, later inviting him to dinner at Watson's Bay where he met David senior, and other members of the family. Allen's impressions were, "David Stead was still very mentally alert, he had strong convictions about things but was not always correct, a very vital sort of person. Thistle? Well, she was not very feminine, a hard character". Thistle never did anything without a pay-off, Allen Strom said.[25]

The Wild Life Preservation Society of Australia had been in temporary hibernation during World War II, but became active again in 1945. By this time David G. Stead was approaching 70 years of age and had not been in good health for some time. He could work at home but was not prepared to attend meetings in the city. David D. Stead accepted temporary appointment to the position of Honorary Secretary of the Society, but was not prepared to carry on indefinitely. Sometime in 1947, when David D. was acting Secretary, Thistle suggested to Allen Strom that he would be a suitable person to take the position of Secretary for the Society. "I might say that Thistle was then at her prime in pushing people around, and as a consequence, I became Honorary Secrerary of WLPSA, whether I liked it or not," Allen Strom said. "As it panned out, it was no great hassle being

Secretary and there were plenty of opportunities to do what I liked doing best, but the big problem was Thistle who kept on pushing people around . . . this could have included me, except that I wouldn't be pushed around, and finally in the 50s told her she better be Secretary and push herself around."[26] For the year 1953–54 Thistle was recorded as Honorary Secretary in place of Allen Strom, and she continued in that position for a number of years, often taking the position of Editor of *Australian Wild Life* as well.

A reasonably successful venture organized by Thistle and Allen Strom was a series of joint camps for students from Balmain and Sydney Teachers' College held at Castlereagh, ten days in February for experience in practice teaching. Sydney Teachers' College had a residential property for student camps at Castlereagh on the banks of the Nepean River. The aim was to spend the first week working out activities for children to undertake in the field, and during the second week bring in children from the local schools while the College students took the role of teachers of field studies. Allan Fox remembers the first camp in 1949; the Balmain students were a tightly knit group under Strom because they were all members of the Caloola Club, but the students from Sydney didn't relate to each other at all. But the learning experience was invaluable, Fox said, because they formed working groups, and were set the task of devising teaching methods for various problems, to be demonstrated to the other students in the course of the day. The groups would alternate with Thistle and Strom, Thistle being more botanically oriented and Strom being animal/ecology oriented. There was friction in the Thistle/Strom team, Allan Fox recalls, but each one contributed an important component for the young College students' field experience.[27] At that 1949 camp Thistle gave each student a set of Adam Forster's coloured plates from her *Wild Flowers of Australia*, an invaluable acquisition in those days when coloured illustrations were rare. The February camps continued successfully for a number of years, probably until 1958 when Allen Strom left Balmain to become Chief Guardian of Fauna and head of the Fauna Protection Panel which administered the

Fauna Protection Act of 1948. Throughout the 1950s David's grandson, John, often accompanied Thistle on her camping trips and he was introduced to ecology at the Castlereagh camps, where the bunkhouse style accommodation was more luxurious than the camps under canvas with Thistle's more adventurous friends.

After her son Fred died in 1938, Ada Stead, who had been living with him, went to Manning Road in Double Bay and shared accommodation with a woman friend. The family used to visit. In 1945 David Darwin married for the second time, Doris Christian (Dorrie), and they settled in Mortdale. David D. was having international success as a photographer, mainly nature photography, and he was well established as a metallurgist. Dorrie was not able to form any close alliance with Thistle, who was always a bit distant with her. Perhaps they had nothing in common; Dorrie had worked in the Accounts Section of the Taxation Department — a wonderful job, she said — and she worked there for fifteen years, including two years after her marriage. Dorrie tried to help Thistle in various ways, but it wasn't easy, she added.[28]

In 1945 two books by Thistle came on to the market. She had been commissioned by William Brooks & Company to do two companion volumes, one she called "*Nature Problems*", a small book of natural history experiments and observations for primary school children, and an accompanying book called "*The Handbook of Nature Study*" for teachers. Both books stressed the value of field work for children, an aspect of teaching and learning in which she was vitally interested.

In May of that year, 1945, Thistle's father died. Her mother continued to be active, a very practical and down-to-earth person. She would go to Boongarre in Watson's Bay each Monday to do the washing, complaining that Thistle was a "dirty puss" who couldn't be trusted to deal with the washing. Perhaps she never forgave Thistle for her allegiance to her father before and after the marriage break-up because she would criticize Thistle to the Stead family and the Watson's Bay neighbours.[29]

Nana Harris also led a very social life. She had her own car and played bridge and golf; she was comfortable in her home at Kirkoswald Avenue in Balmoral, having done very well out of the settlement with her husband.

George Ferguson of Angus and Robertson sent Thistle a copy of Keith McKeown's *Nature in Australia* in April 1948. She replied that she found the book disappointing for two reasons, firstly because there was very little original observation in it, most of it being purely "armchair" labour or perhaps library work; secondly, she found a continual reference to a purpose in nature, or a sort of destiny for every living thing, which no scientist could accept, she said. "These two grave faults prevent me from recommending it seriously to students and prevents me from giving more than a kind of lukewarm approval to it."[30]

In May 1951 Ada Stead died. She had been living alone in a small flat at Double Bay when she had a stroke. Her son, David, took her to his home in Mortdale where she died five weeks later. A month after her death, on June 13, 1951, David G. Stead and Thistle Y. Harris were married at Boongarre. She was 48 and he was 74. All the family members were there, the children riding their bikes around the house in great excitement. On the following Monday Thistle walked into Kath Wall's study at Sydney Teachers' College and said, "I've brought you some cake, I got married in the weekend. Sorry the pieces are so small but the children got into it". "Children?" "Yes, my marriage brought me seventeen grandchildren." Thistle told Merv. Sheldon the news to which he retorted, "Thistle, I hope you didn't rush into this". And she would repeat the story with enthusiasm, because she could laugh at herself.[31] Thistle's colleagues had known she was living with David Stead but their attitude was that it was her affair and it was never discussed. The College staff members had a profound respect for each other, and even a warmth and concern that reflected a relationship almost like a family circle.

During the forties and fifties, at home without paid employment, David Stead continued to write. His many letters to the

Press, not always published, on a diversity of subjects qualified him, he said, as one of the "Prostitutes of the Pen". He worked on his latest book *Sharks and Rays of Australian Seas*, the manuscript completed before his death in 1957. It was published posthumously by Angus and Robertson in 1963. In a change from writing he turned to boatbuilding for his young friends in the neighbourhood. Out of odd scraps of timber, various bits and pieces such as bottle tops, cotton reels and various other odd pieces for ornament, together with a pot of pillarbox red glossy paint he fashioned innumerable little boats which he distributed freely. The local children were constant visitors to exchange their much-used and battered models for new, up-to-date ones while young waterside visitors coerced their parents into taking the beach road to Camp Cove which led them past the boatbuilder's house. Thistle was constantly inveigled into acting as transport officer to deliver to those who came too late in the day to collect their treasures. During these years of their living together, David used to cook the meals during the week, and Thistle cooked at the weekends.

Thistle had a fine collection of records with music from the light classics, Tchaikovsky being one of her favourite composers. But there were two other pastimes she tried to hide because she considered they were vices — reading detective thrillers and the occasional visit to the movies to see a romantic and sentimental film. For about three to four years in the fifties she had a "terrible phase of dressmaking," her niece Carol recalls. Carol and her sister Wendy were teenagers and Thistle would make skirts and blouses for the girls as birthday or Christmas gifts. They were usually run up in haste, and were badly fitting due to obtaining measurements over the phone. "They were terrible," Carol said, "we never had a fitting".

Thistle and David Stead were very much against the holding of animals in zoos, and Taronga Zoo in Sydney came in for its share of criticism. In an article for the magazine *People* in 1951 Thistle, referred to as a "preservationist" said, "The people who manage zoos are often more skilled businessmen or animal-tenders than naturalists. One of the first principles of preservationists is an insistence that wild animals should not become objects of commerce, yet zoos are almost always commercial. Taronga Park Zoo in Sydney, for example, has often been accused by the preservationists of a recurring sin — depending on income derived from the sale of Australian birds and animals."

Another institution that Thistle viewed with disdain was the pseudo-sanctuary. These institutions, often little more than private zoos, call themselves "sanctuaries" either to fool the public or themselves, she said. She added that she found nothing sanctified about capturing wildlife and putting it into cages for a curious public to see. At many such "sanctuaries", the public is invited to buy photographs of themselves fondling koalas which prefer by nature to be on top of a gum tree asleep. "Sanctuaries" are often more concerned with profits than with wildlife preservation, she said.[32] On April 10, 1964 the *Sydney Morning Herald* published a letter written by T. Y. Stead. Among other things she said, writing about Taronga, "Wombats lie concealed in the corners of unadorned narrow pits, and kangaroos lie listlessly on concrete slabs, and monkeys are so far from the point of observation that they can only be discerned as dark blobs against the background". Thistle would surely applaud the changes that have taken place today in the construction and management of facilities at Taronga Zoo and Dubbo Zoo.

In spite of the Stead opposition to zoos, the house at Watson's Bay always had a menagerie of reptiles, birds, guinea pigs, and tanks of fish, but cats and dogs were not encouraged. The guinea pigs bred prolifically and Thistle always managed to find a good home for the countless young ones that came along. The Stead children accepted this happily, and it was not until they were older and wiser that they realized that for someone connected with a Biology Department at a tertiary institution, the "good home" was not necessarily what they had thought. The reptiles were encouraged by David and there was a collection of tortoises, snakes and lizards, plus from time to time various bush "orphans" which people gave to the household because of David Stead's reputation.[33]

Sometime in the mid fifties Thistle set up a large tropical fish tank, primarily as an interest for David, but the children in the house, Dorothy, John and Robert, became the curators. A lot of money was spent on these tropical fish as the acclimatization of the fish was very much trial and error with random acquisition and watching to see which ones were attacking the others. The tank needed to be heated and on one occasion there was a "problem" with the thermostat and the whole tankful of fish became overheated and died.[34]

One of Thistle's publications, *The Teaching of Nature Study*, published in 1954 by the Australian Council for Educational Research (ACER), was the result of her being awarded a research grant by the ACER in 1950. In 1951 she undertook research into the teaching and learning of nature study in 42 primary schools located all over New South Wales, presenting a lengthy report to the ACER in December 1952. While the report indicated that there had been an honest attempt to involve a large sample of schools, teachers, and children in the project, the study, as a piece of empirical research, had some serious flaws. The report also was of average quality, reflecting the defects of the study itself.

The study was entitled "An Investigation into the Teaching of Nature Study in Primary Schools (Third and Fourth Grades) with some Evidence of its Basic Value in the Training of Young Children". In selecting the 42 schools of 8 different types (city, rural, one-teacher, central, etc.) Thistle based her choice on schools with interested and co-operative teachers. The undertaking, Thistle herself stated, was a considerable one, involving the active co-operation of principals, teachers-in-charge of small schools, some eighty classroom teachers, and more than two thousand children from schools of the New South Wales Education Department. She visited all the schools either before or during the course of the project. The study involved a pre-test for the children on knowledge, principles and attitudes concerning natural history (nature study), a six-month concentrated teaching programme (to include soil, plants, insects, birds, and general studies), followed by a post-test in November of 1951. Thistle marked all the papers herself to maintain a standard of assessment. A serious flaw in the study was a failure to express the hypotheses in a form which could be tested; rather, the four hypotheses as stated in the Introduction, expressed beliefs or recommendations already decided. For example, Hypothesis (i) states:

> Nature Study should be a basic subject in both primary and lower primary schools, round (sic) which the fundamental subjects recognized at present (namely English, Arithmetic, Social Studies) should be built.

Hypothesis (ii) states:

> Nature Study is not regarded as such a basic subject at present in these grades in New South Wales partly because . . . (here follows four of her reasons based on previous observation).

As part of standard procedure for a study of this type, Thistle carried out a preliminary test at Teachers' College Camp in February 1951, using children from Warragamba Public School. In this preliminary test, there was a group of 34 children who attended camp and received instruction between the pre- and post-test, and a group of 34 children who did not attend camp or receive instruction between the pre- and post-tests. However, the children's ages for this preliminary test ranged from 8.2 years to 13 years, with most children being 10 or 11, not a satisfactory indicator for an investigation where the children to be tested would range from 8–9 years. Teachers in the actual study were to later comment on the unsuitability of the final test for 8–9 year olds. This preliminary investigation did not offer any recommendations.

However, for the actual study, it was wisely decided to arrange the pre- and post-tests in three sections, instead of one as early planned; it was also decided to make the tests pictorial as far as possible, to require the shortest possible answers, and to set no time limit. Thistle advised the teachers to give each of the three tests on a different day. Test A aimed to test knowledge of Natural History. Test B aimed to test principles involved in Natural History. Test C aimed to test attitudes to Natural History in general, and to conservation of fauna and flora.

It is clear from comments made by the teachers that the preliminary investigation should have been carried out on these three tests, on 8–9 year olds, and the tests subsequently modified. Teachers said:

> "Too long and would have been better for older pupils."
>
> "Pupils floundered in Test C, mainly due to lack of literary ability."
>
> "Test B was difficult for poor readers."
>
> "Children could not understand some questions due to the phrasing used."
>
> "Better results may have been obtained from 6th class for many questions seemed to be beyond the conception of 8-year olds."
>
> "The space left for answers was too small for third class writing."

On a more positive note one teacher said:

> "The results of the tests show noticeable improvement in all phases of work. It is evident that the children have benefited from the project. I think it is equally true that the teachers participating also derived much benefit."

As part of the research project Thistle herself undertook a brief teaching experience with a fourth class of 43 boys in a suburban school, and admitted that she learned "a tremendous amount". It involved Thistle taking the class for one hour three mornings a week over a period of six weeks in the summertime, just before the Christmas holidays in 1951, the emphasis to be on field work, mainly in the school grounds.

Thistle also undertook observations and a few teaching ventures during the course of the study, this restricted to two 10-day periods in vacation times and a six-week period of special leave from her work at Teachers' College in the latter half of 1951. She recounts two interesting teaching experiences:

> "At one school I said, 'How big is the class?' and was told 40. So I prepared 40 sets of apparatus. When I reached the school, I walked into a room with two children sitting at every single desk, and with a crowd of children standing along the back wall. The class teacher said, 'Excuse the crowded conditions. We thought it was a pity for the classes to miss the opportunity, so they sent in two more classes.'
>
> "At another school the teacher had elected to have a lesson on the setting up of an aquarium. I assembled the material, except the aquarium itself, of course, for demonstrating this to the class. At the school I said, 'All right, I've got all my equipment here, now bring out your aquarium'. 'Oh,' said the teacher, 'we don't have an aquarium, we just thought it would be nice to know how to set one up if we ever got the chance of getting one'."[35]

Another part of the research study involved looking at children's conceptions of the natural world. A questionnaire of 40 questions was given to those children in the 42 schools of the sample population. Questions such as these were asked:

Q.8 What makes clouds?

Q.15 Can a plant feel the wind blowing?

Q.27 Do flies have mothers and fathers?

Q.35 Are fish ever sad? What makes you think this?

Q.38 Why do birds have feathers and not fur?

This was an interesting aspect of the study, but Thistle's analysis of the children's answers was heavily dependent on the ideas of Jean Piaget. Piaget was an educational psychologist whose ideas on developmental stages in a child's cognitive growth were popular at the time.

The section on General Conclusions at the end of the report was somewhat disappointing because it was a discussion of ideas, her own and a number of authors quoted in detail, rather than a correspondence between the hypotheses being tested and the conclusions to be drawn from the study alone. But then the hypotheses themselves were poorly stated. Again, the section on "Some Constructive Suggestions" is 95% or more taken from other authors.

There was good discussion by Thistle on pages 146 ff of problems revealed by the project with the teaching of nature study, with respect to the teachers themselves. Some useful suggestions were given concerning problem-solving and field work for children in primary school.

The ACER publication written by Thistle and published in 1954, was a much more robust and comprehensive document than the report discussed above. It was based on the 1951 study, but drew also on Thistle's experiences with higher primary level children and her long experience with Nature Study in the schools. The book was a useful guide for teachers, with suggestions on how to organize field work, how to select topics for study in the field, outlines of lesson plans in which field work may be used, conducting experiments, making records, writing programmes, and guiding principles for conservation. The book is to be regarded as a worthwhile attempt to encourage teachers lacking confidence in the sciences to see the value of problem-solving, first-hand observation, and studies beyond the classroom, at the primary level. In contrast to the research report, it is a practical, well-written, useful book.

Thistle Harris drew considerably on the knowledge and advice of other enthusiasts and specialists in the compilation of her books on Australian plants for the garden. Her great skill was in being able to know what information was needed, collect that information, and then present it in an acceptable and meaningful form for the non-specialist and garden lover. She was also improving her skills in photography during the fifties. John Stead believes her photographic expertise probably peaked in the sixties, and around this time she amassed thousands of slides of flora which she used to illustrate talks and publications, and as memory joggers when she was writing. She spent thousands of pounds/dollars on photographic equipment, always trying to get something better for the close-up work. John's first good camera was a Thistle cast-off, and he also profited from her accomplishments as a tutor in matters photographic. She had a darkroom built in the house at Watson's Bay and for many years they processed their own prints, which enabled Thistle to edit her photos and only have those negatives processed which adequately covered her subject matter.[36]

The first book for the garden by Thistle was published in 1953, a more comprehensive work than any previously published book on Australian garden flora, although in later years, Thistle herself would criticize the book for its many failings. But as a pioneer, breaking new ground, it filled a gap for the garden enthusiast; as Thistle said, "Looking at this book today, I look at it rather with horror, because I realize that I knew very little about the cultivation of Australian plants at this time, and there were very, very few that were available on the market. I knew this, I knew that I didn't know very much, but I felt that this simply had to be done, that the Australian people needed some guide in this sort of field and even if it wasn't a very good guide, it was something, and I do think this book helped to set the culture of Australian plants on a good footing."[37] In embarking on this project, Thistle displayed some courage (or was it bravado?) but more importantly, the attitude of, "Here's a need to be filled, let's do it".

While the book reviews applauded this 1953 publication as being among the most comprehensive and useful yet produced, they did draw attention to some of its weaknesses. "Being a pioneering book in this particular field one expects to find imperfections," the *Agricultural Gazette* reviewer wrote in October 1953. "Soil and habitat requirements for a number of plants are hardly correct . . . The selection of species for general treatment seems rather haphazard." A review found among Thistle's papers, acknowledged only by initials K. M. and J. G. is very critical of this 1953 book.

"... the subject matter is neither well balanced nor well arranged and many of the statements show an inadequate knowledge of the plants concerned ... The examples chosen in the chapter on the naming of plants will add to, rather than eliminate, confusion in the reader's mind ... Many of the individual species are inaccurately described ... Inaccuracies also extend to the naming of species figured, the appendix of derivations of scientific names and to many of the botanical names throughout the book. In spite of these defects Miss Harris has assembled a great deal of information which will be of interest to the general reader and will perhaps help to stimulate an appreciation of our native plants as garden subjects."

It has already been pointed out that Thistle Harris owed a great debt to the particular kind of people she met throughout her life, people who acted as mentors or catalysts or sources of information. In *Australian Plants for the Garden (1953), A Handbook on the Cultivation of Australian Trees, Shrubs, Other Flowering Plants and Ferns,* Thistle acknowledged her debt to Ivo C. Hammet of Victoria, "for it was largely due to his enthusiasm for his own native garden and to my many pleasant rambles in it that I conceived the idea of trying to help others to emulate it". Once motivated, she called on the assistance of others in herbaria around Australia, in the Forestry Departments of various States, and in public libraries; H. T. Reeves of Melbourne put his collection of flower studies at her disposal; Edna Walling, landscape architect of Victoria, helped in the design of garden plans; George W. Althofer of Nindethana Nursery, Dripstone, New South Wales, gave much needed assistance in the technique of cultivation. In the compilation of this work, as in others to follow, Thistle Harris took on the role of an educator, rather than a researcher, except in the sense of searching out the information from others.

An example of Thistle's dependence on expertise from others is seen in the garden plans she included in the back of the 1953 book. Peter Watts, Director, Historic Houses Trust, an authority on Edna Walling and her work, believes that the basic plans in

Thistle's book have been drawn by Edna Walling, obviously done specifically for the book as rough drawings or sketches, and modified and redrawn by Norman Weekes, architect, of Sydney. The plans are very much in the style of Edna Walling design, an Edna Walling concept, with Thistle's ideas for the plants superimposed.[38] Soon after the first edition of *Wild Flowers of Australia* in 1938, Edna drove from Melbourne to Sydney in one day with the express purpose of meeting Thistle. During the 1930s Edna had made her name as a landscape designer around Melbourne for many of the city's rich middle-belt suburbs. In those days she made more use of exotic plants, but in the 1950s she almost totally rejected the use of exotic plants in favour of native species. She died in 1973. Thistle did acknowledge the help of Edna Walling and Norman Weekes in the compilation of her garden plans.

C. T. White, Government Botanist in Brisbane, wrote in his Foreword to the 1953 publication that "few people have done as much and probably none more than Thistle Harris to popularize the study of our wild flowers". White was prophetic in this same Foreword as he stated, "I firmly believe that we are only on the fringe of the utilization of our Australian plants as garden subjects, and we may yet see many of them vying with the more usual annuals as bedding and border plants and as subjects for the herbaceous border".

One pioneering aspect of the 1953 book, and one to be applauded, was a list of derivations of scientific names in the Appendix, a considerable help to those who could be deterred by botanical names, and no doubt a fascination for those who revel in the classical languages and who appreciate the association of plants with places or personal names of botanists. This inclusion reveals the academic botanist behind the nature lover, and a similar list in a later 1980 publication *(Trees)* avoided the errors of this earlier work.

Thistle's 1953 book on the cultivation of native plants had a prolonged and frustrating incubation period. In January 1947 she wrote to George Ferguson re "the proposed book on cultivation of native plants", saying she would like to get as much completed as possible during the forthcoming six weeks. Cyril White, Government Botanist of Queensland, had offered to go through the manuscript with her if she travelled to Brisbane. "He is the most competent man in this particular field in Australia . . . He is a personal friend of mine and is interested in the general aspect of the work. It would be most convenient for me to fly up in the first week of February." It was not until January 1948 that she went to Brisbane to see Cyril White.

In July 1947 Thistle was writing to Ferguson asking him to send a photographer to the property of Mr G. W. Althofer at Dripstone (near Wellington, New South Wales) to get suitable photographs for the book. She also asked for the photographer to travel to Melbourne for the same purpose. Ferguson replied that Mr Davis, in charge of art work and photography at Angus and Robertson, would go to Dripstone, but Melbourne was out of the question. However, in the meantime, Mr R. G. Kimbell of Wellington, sent three coloured transparencies of native plants to George Ferguson, saying he would send more in the future.

Thistle wrote to Ferguson in September 1949, complaining about delays in the Garden book, which was supposed to come out early in 1950. On July 19, 1950 Angus and Robertson finally sent her the agreement for her signature, but in December of that year arrangements were still being made for photography work. In April 1951 Ferguson wrote to Thistle to say how much work was needed on the manuscript, adding that had they realized just how much work was going to be necessary on their part, they would never have accepted the manuscript in its original form. "Having regard to the costs already incurred I am wondering whether you would agree to varying the agreement which has already been signed to permit us to pay you a royalty of 5% on the first thousand copies and thereafter 10% as arranged." Thistle replied to this letter to say she could not agree to cut the royalty, especially since there had been such a long delay, adding that the manuscript had been in their hands for well over five years.

A full page typed memo, no date, no signature, setting out the reasons for lengthy delays in preparing the manuscript for print, is in the Angus and Robertson papers in the Mitchell Library. It states that the original copy was the most confused this editor had ever seen. It was necessary to retype the manuscript, but before this could be done, there was the difficult job of making clear, for the typist, in what order the extensive hand-written additions were to be typed. These hand-written additions in some cases were so illegible that the author had to be consulted. Many queries became obvious after the retyping and these too had to be referred to the author. Finally, the editor decided to check the many cross-references; so many serious inconsistencies became apparent that it was obvious only an expert would know which was the correct statement of two conflicting ones. The manuscript was returned to the author, but once again, owing to the extensive corrections, it was necessary for the editorial staff to go right through it again, clarifying the corrections and checking the final materials.[39] The book was finally published in June 1953, and in July 1953 Thistle wrote to Mr W. G. Cousins of Angus and Robertson to say she was to be interviewed on the ABC by Frank Legge on the evening of Saturday, July 18, adding, "I think the book will be generally well received".

One is led to speculate that Angus and Robertson must have been confident of their market for a book of this type when they were prepared to continue working on a less than satisfactory manuscript presented to them by an impatient and careless author.

Thistle's typed manuscripts were the product of her old portable typewriter that always seemed to need a new ribbon, and had been pounded by the Stead children. It missed certain letters, and the letters that did appear seemed to go up or down half a line at will. Her handwriting was also quite erratic, John Stead recalls, and consequently any communication from Thistle became a case of breaking the code. The manuscripts for two of Thistle's later books published in 1977 and 1980 were typed by John's wife, Patricia, who was able to sit down with Thistle and slowly decipher her handwriting.[40]

Throughout these busy and turbulent times Thistle continued her trips around the country, usually in her own car. Her bad driving was legendary. Allen Strom recalls that after the war, in 1946, she acquired a new car and was told not to rev it too much, so she drove it in second gear all the time. "She had no idea about the mechanics of cars."[41] A staff member at College said to Kath Wall one day, "Have you ever been driving with Thistle?" "Yes, why?" "Well, if you go out driving with Thistle, and she sees a tree she wants to look at, she drives across the field and up to the tree!"[42]

Thistle became very friendly with several mature age women students who were part of a special intake in September 1950 to help overcome the shortage of teachers. One of these students was Connie Burns, who had captured Thistle's eye because of her interest in Thistle's natural science course. Connie attended two of the natural history camps at Castlereagh, one in February 1951 and another in February 1952. But Connie and a fellow student became great company for Thistle on two of her trips to visit country schools for the ACER research project. The first trip was out west in May 1951, and the second in August 1951, before the school vacation, up the north coast almost to the border and back through New England. It was vacation time for the two College students, but they learned an enormous amount, not only about research in a school environment, but about the various environments through which they passed. Connie Burns also accompanied Thistle on three of her annual pilgrimages to Kosciuszko.

In one of the 1955 school vacations Connie Burns ran an inservice course on primary school natural history for Thistle, at Sydney Teachers' College. Thistle organized the course, but Connie provided the input and ran the workshop. When approached once again by Thistle, Connie had to refuse because she was busy with studies at University. Connie Burns and her friend, Doreen Clark, were to be of great support to Thistle at a later date when they worked as volunteers at Wirrimbirra Sanctuary, and took on the demanding job of co-editors of the newsletter *Wildlife Research News*.

Another book had its genesis in the 1950s. In July 1949, as President of the Wild Life Preservation Society, Thistle had written to Angus and Robertson to suggest the WLPS would produce as a joint effort a "good handbook for bushwalkers — Naturecraft for the Bushwalker". George Ferguson replied that Angus and Robertson were interested if Thistle would be the editor, and Angus and Robertson would pay 10% royalty to the Society. In October 1955 Ferguson was writing to say *Naturecraft in Australia* (its revised title) could not be ready for Christmas. It had been another long "winter of our discontent."

Chapter 4 Notes

1. de Berg Tapes, Tape 574 NL, January 28, 1972, Side 1, p. 8.
2. Notes from Adam Forster file in National Library, Canberra.
3. Information on Adam Forster's background from great-granddaughter Elaine Coldham.
4. Pers. comm. 3.4.97.
5. Examination of paintings at Harper Collins 3.4.97. *Grevillea punicea* is now *G. speciosa.*
6. de Berg Tapes, Side 1, p. 9.
7. ibid. p. 10.
8. ibid.
9. Angus & Robertson correspondence, ML MSS 3269, Vol. 324.
10. *New Horizons in Education,* April 1943, p. 24.
11. "A Thistle Among the Wildflowers", article in *People,* June 20, 1951.
12. *New Horizons in Education*, April 1943, p. 25.
13. *Religion in Education*, E. O'Brien *et al.* 1943, p. 3.
14. ibid. p. 16, Very Rev. J. C. Thompson, Rector, St John's College, University of Sydney.
15. ibid. p. 25.
16. ibid. pp. 41, 43, 50.
17. ibid. Preface.

18. *New Horizons in Education*, October 1945, p. 30–31.
19. ibid. Vol. 5, No. 1, May 1947.
20. *Education*, Vol. 23, No. 9, July 25, 1942, p. 270.
21. C. Hinder, pers. comm. 12.11.96.
22. J. Stead, pers. comm. 14.7.97.
23. A. Strom, pers. comm. 3.1.96.
24. A. Fox, pers. comm. 17.3.97.
25. A. Strom, pers. comm. 2.1.97.
26. A. Strom, 15.1.1986, letter to I. Bowden.
27. A. Fox, pers. comm. 17.3.97.
28. David and Doris Stead, pers. comm. 2.11.96.
29. D. Basili, pers. comm. 29.1.97 and C. Hinder, pers. comm. 30.6.97.
30. Angus & Robertson correspondence, ML MSS 3269, Vol. 324.
31. K. Wall, pers. comm. 25.6.96.
32. Article in *People*, June 20, 1951.
33. J. Stead, pers. comm. 14.7.97.
34. ibid.
35. de Berg Tape, Side 1, p. 7.
36. J. Stead, pers. comm. 14.7.97.
37. de Berg Tape, Side 1, p. 11–12.
38. P. Watts, pers. comm. 4.4.97.
39. Angus & Robertson file, ML MSS 3269, Vol. 324.
40. J. Stead, pers. comm. 14.7.97.
41. A. Strom, pers. comm. 2.1.97.
42. K. Wall, pers. comm. 25.6.96.

CHAPTER 5

Nature's Advocate — 1955–1963

"The natural world — it is our heritage and our responsibility."

T. Y. Stead

HOWEVER influential Thistle's books may have been, it was in the classroom, the laboratory and in the field with students that she made a lasting impression. Her basic philosophy about the teaching of natural history is summarized in the passage below:

> "Australia offers to the student of nature a vast and amazing array of objects for study, not least among them being the plants. The more one comes to know them, the more one finds to learn and the more absorbing does the study become. Perhaps the need to inspire students with some appreciation of the bush has taught me a greater love and respect for it. I believe firmly that people learn most readily if they come into direct contact with the object of their study and so, as a teacher of biology I have always taken students out into the bush as much as possible — sometimes for an hour or two, sometimes for much longer periods. This is an enjoyable way of teaching and, for the teacher also, of learning, for one cannot explore the bush without learning something each trip."[1]

The reaction of students and colleagues to Thistle's teaching varied, but it was never one of indifference. "My recollections of Thistle Harris are good ones, but nothing remarkable. I found her a thorough and caring mentor who guided us into teaching in a practical way. I was ever grateful for her preparation when I took up my first posting. I was well able to cope because of her preparation. I was aware of her independent spirit, that she trod on academic toes, that she had socialistic leanings and was keen to see native plants in gardens. I was told, but mostly by innuendo, that she relied heavily on other people's expertise and that her books reflected others' work to a greater degree than her own." So wrote Doreen, a science graduate in the Diploma of Education at Sydney Teachers' College in 1950.[2] Another science graduate in the class of 1950 recalls Thistle's masterly strategy at the beginning of the course in Biology Teaching Method: she asked the university graduates to complete the 1949 Leaving Certificate paper in Biology. What a cinch, these students of science believed. Perhaps Thistle erred on the side of strict marking, but it was a salutary lesson when the returned papers showed an average mark of 68%.

"Energetic, enthusiastic and inspirational — these are the attributes of a good lecturer," said Patricia, "and in 1951 Thistle was none of these. She slouched in and slumped into a chair, or perched on a stool, not fashionably dressed, no lady, no example, no role model for young trainee teachers. Did she have reasons for this? Her mind, perhaps, was on other things, her ACER project and getting married. She was more of a naturalist than a scientist."[3]

"Lecturers like Thistle Harris tended to stand out as being prepared to accept her students as young adults or mature students and treating them not as docile persons to be told what to think and do next," said Mary, a student in the Diploma of Education at the College in 1950. Mary had completed a three-year course for a Bachelor of Arts degree at Sydney University; students training to teach in high schools at that time had to select three subject methods for their future teaching, and Mary chose Biology for her third subject, thus coming into contact with Thistle Harris. Later, when Mary was "practice teaching" — the term used for a student's first efforts in the classroom — Thistle came to supervise at Hornsby Girls' High School. Mary found Thistle's interest and support in her "rather pitiful efforts to present Biology to a junior class very comforting" and at the same time her analysis of Mary's mistakes was clearly defined and positive. The way schools were structured at that time meant that it was unlikely that Mary would ever teach Biology but this did not lead to Thistle giving any less of her time to that individual student teacher.[4]

Mary has provided another aspect of Thistle's contribution to College life: "I was closely involved with leftwing radical politics, being a member of the Sydney University Labor Club in 1950. There was a strong force of radical students in 1950, including some exservice students, and this combination produced at times a rather volatile atmosphere in the Teachers' College when authority was challenged on a number of issues related to the administration of the College. Thistle Harris, although not involved in these clashes, provided guidance and understanding in her lectures with great emphasis on questions of humanity, the need to have a philosophy of education and the importance of developing the minds of students who were to be our responsibility in the schools. Wonderfully knowledgeable, she was nevertheless easy to approach, her eyes sparkling with enthusiasm for her subject and for teaching, which she saw as having high value and prestige as a profession."[5]

Thistle lived in a world of her own, Isa (Robson) Cohen, a colleague at the College reports. She was a loner, even though she mingled easily; she didn't need anyone's approval or disapproval for she was single-minded, and once having made up her mind on a course of action she pursued it with determination, certain of her own integrity. As a loner she was very self-sufficient and did not have one special close friend among the staff. But in those days the College fraternity was one closely-knit family and a number of staff members received an invitation to afternoon tea at Boongarre. Thistle may have been a loner, and perhaps a little self-centred, but she was always interested in other people's problems, especially if they were questioning things. It was her idea to rally some staff members to take exam supervision for a colleague whose father was dying, hers the first name on the list. Perhaps self-centred, but Thistle was not selfish. "Selfishness" said Oscar Wilde, "is not living as one wishes to live. It is asking others to live as one wishes to live".[6]

Staff meetings at Sydney Teachers' College were an opportunity for Thistle to play a part in policy making, and she was quite fearless in expressing her opinions, firm yet courteous. If she dismissed another's opinion it was never done with malice but with a little derisive laugh. However, she was always careful to avoid discussion on religion or politics and did not intrude on other people's convictions.[7]

A number of Thistle's colleagues at the College recall that she had loads of physical vitality, yet she did admit to one of them that sometimes she had to stay in bed for a couple of days to restore her strength, especially near the end of the year.[8] A member of the Science staff said she was fun to be with, never pompous or self-important, and prone to tell jokes against herself. Yet she agrees Thistle was never a great scientist, her organization was not always up to scratch, and all sorts of things happened.[9] On the positive side, however, she was a pioneer in the battle to get Nature Study in the primary school recognized as a scientific pursuit. Even in the colleges this was a debased subject and it was through Thistle's efforts that it became more respected. "She was an enchanting teacher of small children."[10] Allen Strom was of the opinion that Thistle should never have been an educator, that is, a teacher at the coal face, because she didn't have clear goals, or programme her work. She was much better as a source of information, through her books, he added.[11]

Yet Joyce Wylie, another colleague at Sydney Teachers' College, saw her as an excellent teacher, very forthright, devoted and dedicated to her area of expertise, a keen conservationist, and not afraid of hard work. She had a good sense of humour, and didn't suffer fools gladly. Joyce was always happy to leave work and get on the tram with Thistle to go home.[12]

Brian Morley spent two years as a junior lecturer at Sydney Teachers' College in the science department, working under Merv Sheldon in 1952 and 1953; Brian had completed his Diploma of

Education in 1951 after a Science (Honours) degree at Sydney University. Merv and Thistle each had their own laboratory, and each lab had a freshwater aquarium with goldfish. One day Thistle asked to borrow Merv's aquarium, and kept it for some time. Merv eventually asked Brian to retrieve his aquarium; Brian went to Thistle's laboratory and saw two aquaria there, one with dead fish and one with live fish. When Brian said, Merv wants his goldfish, Thistle said, his are the dead ones! Brian said Thistle was well ahead of her time in sex education and conservation education but he refused to be drawn into any discussion about her standards.[13]

A student at Sydney Teachers' College in 1958 tells the story of following Thistle up the stairs one day, to be overtaken by two brash young male students. Said one to the other, "Don't thistles have thick stalks!" to which Thistle Y. Harris responded with a dirty look.

A fine acknowledgement to the influence of Thistle Harris on a young student is given by Lionel Gilbert in his book, *The Royal Botanic Gardens, Sydney: A History 1816–1985*. He wrote, "Probably I should never have contemplated this project had I not experienced the good fortune to be introduced to the study of botany by Miss Thistle Y. Harris (later Mrs David Stead) at Sydney Teachers' College in 1942. Nor have I forgotten her kindness in writing a biology course for a class of one to enable an essential 'option study' to be completed by external study during the unsettled years of 1943–45."[14] Lionel became a lecturer at the University of New England, Armidale, and at the launch of his book on the Botanic Gardens in 1986 he referred again to his good fortune in selecting Biology as an optional field of study. "I had the pleasure of Miss Thistle Harris's enlightening lectures in classroom, laboratory and in the field, for six hours a week, and sometimes at weekends as well, while a new world was revealed to us in the College and University grounds, in the wilds of the Woronora River Valley, at Longreef and even in Sydney Harbour."[15] Thistle was to say later about Lionel Gilbert, "Best student I ever had and he had to go and do history".[16] However, Thistle's influence was there because Lionel did make sorties into botanical history. Thistle's book *Wild Flowers of Australia* was the first introduction to Australian flora for the young Judith Steanes, and when Judith studied Botany at the University of Sydney in the 1950s, the book became a useful aid, as it had been for many other students. Mrs Stead was the Lecturer in Biology Method when Judith did her Diploma of Education in 1956; her memories recall a frank and honest woman, but above all a practical person. Her stress on scientific method and critical thinking reinforced Judith's belief that herein lay the essence of science, not only the learning of facts. In 1962 Thistle contacted Judith to invite her to join the staff at Sydney Teachers' College as a Lecturer in Biology, and so the relationship changed again — "she was a very free and easy Head of Department, practising her belief that you choose the right people and let them get on with it".[17]

Athol Berglund, now retired from his lecturing position at the University of Technology Sydney, Kuring-gai Campus, used to give demonstration lessons for Thistle's students at North Newtown Demonstration School. Athol has some vivid memories of his association with Thistle over several years.

> "In 1958 I was appointed to North Newtown Demonstration School. The school had two small patches of garden and staff were asked for volunteers to run these. Since I was the only keen gardener, I got the job. The school shared grounds with the North Newtown Annexe of Sydney Teachers' College and the College sections had flourishing native gardens supervised by Thistle Stead.
>
> "My first meeting with Thistle occurred when some children and I were digging in preparation for planting. Suddenly there was a voice: 'What are you going to plant in it?' it was Thistle, although at that stage she didn't know who I was and I didn't know who she was, although I'd been using *Wild Flowers of Australia* since my College days in 1950–51."

Athol recalls that Thistle was to supply all the plants for the garden patch, and was a marvellous help with details of culture and planting information. She frequently talked with the children when they were gardening but only ever about plants — no idle chatter — no unrelated matters.

Athol's recollections continue:

> "Over the next two and a half years, I talked frequently with her and demonstrated lessons for her College classes. Our conversation was always about science, gardening, and science lessons. She rarely used names for people and I sometimes used to wonder if she really knew who I was.
>
> "The school staff in general dreaded giving lessons for Thistle. She was easily the least favourite College lecturer and many went to great lengths to avoid demonstrations for her. Staff complained that her appraisal of lessons was invariably uncomplimentary and abrupt. Others accused her of leaving classrooms without comment or thanks. Many staff criticized her manner and her own teaching skill.
>
> "On the other hand, I found her very specific in what she wanted and incredibly involved with the matter as well as the process of lessons. I suspect that none of us really knew enough science for the kind of lessons she wanted . . . at all costs she wanted accurate scientific material and no emotive overtones or interpretations.
>
> "In retrospect, I found her challenging and exacting. I certainly remember the times I worked for her with great clarity. She was a strong woman with a pioneering spirit and a real mission in life. In the 1950s in Australia, women professionals needed not only strength but uncompromising toughness. I think Thistle Stead had these qualities." [18]

It is significant that Athol Berglund and Lionel Gilbert, two very perceptive people who themselves became competent educators, were strong in their positive comments about Thistle's contribution to education.

Thistle depended greatly on the advice she received from the staff of the Botanic Gardens, especially in writing and revising her books. Lawrie Johnson, Emeritus Director of the Royal Botanic Gardens in Sydney, said that Thistle had the respect of the professional botanists at the Herbarium; she was not a scientific contributor, she was an amateur, he said, but she was an earnest hard worker who kept in touch with the botanists at the Herbarium, not just to find out information, but to tell them of things she had observed — it was a two-way thing. Lawrie said, "It was an uphill battle in the fifties and sixties, and earlier, to get anyone in authority, or in business, or among the general public, to take an interest in environmental matters, but Thistle was steadfast in fighting for conservation, especially in educating people about the environment. She knew a lot of plants in the field, but I guess you could say her botanical expertise was derivative. She was a good person to deal with, always good at keeping up with the latest in classification, although like a great many people she did object to the idea of breaking up the Eucalypts. She was a great battler for conservation."[19]

Family members claim that Thistle was a member of the Communist Party, but other people who knew her in the fifties say this was unlikely because of her extreme individualism and reluctance to toe any party line. One family member recalls that some time in the early to mid 1950s, Thistle resigned from the Communist Party, disgusted with the ideas of Lysenko.[20] Trofim Denisovich Lysenko was the most important biologist in Russia from the mid-1930s to the late 1950s. He was the leader of the Soviet school of genetics that opposed Mendel's laws of inheritance and maintained that acquired characteristics are inherited. Joseph Stalin supported Lysenko's theories for improving farm production because they promised quick success, but the theories proved faulty, and later Russian leaders blamed them for slowing the growth of the country's agriculture. It is claimed that Lysenko held back Russian research in genetics (the science of heredity) for more than 20 years. After the death of the Soviet leader Joseph Stalin in 1953, Lysenko was strongly criticized and his influence diminished, his ideas rejected by modern biologists.

Mary Reynolds, mentioned earlier as one of Thistle's students in 1950, continued to be involved with left wing politics throughout the 1950s and into the early 1960s. "For us at that time there was a profound belief that society could be more equal and just for all; this belief drove our approach with a deep sense of commitment. Tied in with that determination was a genuine dedication to the Peace movement and the Banning of Nuclear Weapons — the latter seen as an awful and terrifying threat to the world after the ending of World War II. Regular participation in the Hiroshima Day Marches around August 6–8, other Ban the Bomb activities and the International Peace and Disarmament Movement led me to realize that Thistle Harris also had deep concern for these issues. She was often present at such functions and at times took an important role in them; I recall her particularly at meetings at the Rushcutters' Bay Stadium, a venue used for gatherings after the Hiroshima Day Marches."[21]

Mary has pointed out that mistakenly at that time these movements for peace were too closely linked with the U.S.S.R. and its policies. Nevertheless, this did not mean that the majority of those who participated were not possessed of a deep belief in these causes for, at the time of cold war politics, war and the use of nuclear weapons were a constant threat. "We were all subjected to attack and vilification," Mary has said, "especially bad during the McCarthy period in the U.S.A. Thistle Harris, being a person of standing in the community, had to endure this, but she never wavered from her basic beliefs. She remained strong, always retaining her integrity, courage and determination."[22]

Conservation for Thistle was of equal importance in the struggle to make a better world, Mary Reynolds believes. In the fifties and sixties the environment and conservation as concerns were seen as almost absurd and ridiculous. Among left wing people these matters were viewed as anti-working class and therefore quite unacceptable. Mary and her husband, Ellis, developed a strong interest in Conservation at this time and were members of a local residents group in North Hornsby. In a small leafy street in North Hornsby a group of Eucalypts was under threat because they occupied the centre of the road, an unusual position it is granted, and some saw them as a danger. Thistle was asked to examine the trees to provide a scientific assessment of their position and quality. Her favourable report was largely responsible for the trees remaining *in situ.*[23]

In 1956 the book produced under the auspices of the Wild Life Preservation Society, *Naturecraft in Australia* was published by Angus and Robertson with Thistle as editor. It was planned as a guide for the nature-lover, the bushwalker, the student and the teacher, with chapters on geology and the formation of rock and landscape, animals of the bush and of the seashore, flowering plants, fungi, ferns and mosses. For those interested in spending nights as well as days out of doors, there was a section on the stars and also a section with rules on campercraft. "Bushwalkers and others will also learn how they can help to preserve the bushland that means so much to them." A second revised edition was released in 1967 and to this had been added chapters on Nature Conservation, Nature Protection and the Conservationist, and National Parks and Reserves. Vincent Serventy, who helped in the editing, said the book didn't sell so well, it sold steadily but slowly, probably due to too much competition, he added.[24] But there would be few books of this kind on the market, specifically aimed at the bushwalker and conservationist; perhaps the book was ahead of its time, with general environmental awareness only becoming a phenomenon in the seventies.

In February 1957, Christina Stead had written to Doris and David Stead to say, "I had a letter from Thistle, poor girl, about Dad's health, she says not to be sorry for her — of course I am".[25] David had been in poor health since 1955 when he had an operation for prostate cancer, from which he never really recovered. He took to his bed and Thistle nursed him with loving care, sparing him no attention that she could give when not at work. Ivy Stead was a great support to Thistle in the nursing, and Doris Stead often went to Watson's Bay to relieve Thistle and Ivy. He eventually died on August 2, 1957, aged 80. Thistle was so devastated she refused to go to the funeral, she was angry with him for dying, having tried to prolong his life in the previous weeks with blood transfusions. From being a person who was easy to live with, Thistle became distant from the family, her trips away became more frequent.[26] The family persuaded Thistle that she would not have the time to look after all of David's accumulated treasures, so they were dispersed, much of it, such as native art, to the Australian Museum. His death passed little noticed for a man who had contributed so much to fisheries and conservation, but in March 1966 the New South Wales Surveyor-General approved the naming of a peak in the Central Blue Mountains, "Mount David G. Stead". It is shown on most maps simply as Mount Stead.

In the months leading up to David Stead's death, Thistle had been conferring with a friend in Victoria, Mr A. J. Swaby of Melbourne. For several years, from May 1954, Mr Swaby wrote a column "Know Your Natives" in the popular gardening magazine, *Your Garden*. He had a deep conviction that Australian plants deserved a prominent place in Australian horticulture, and these excellent articles attracted a great deal of interest from persons scattered across Australia. The enthusiastic response to his articles inspired in him the idea of forming a Society, and he announced the idea in the April 1956 issue of *Your Garden*. In December, Mr Swaby reported "Applicants to share in the foundation of an association of 'Australian Growers of Australians' exceeded expectations many times over". The inaugural meeting of the Society was held in the Horticultural Hall, Victoria Street, Melbourne, on March 12, 1957, at 8.00 p.m. Over 400 growers rallied to the formation of an association to be known as Australian Growers of Australian Plants; visitors from other States added to the numbers present. Mr Ivo C. Hammet, another friend of Thistle's, was elected President, and Mr Swaby became Organizing Secretary. Along with eight others, Thistle was elected an honorary life member.[27]

Swaby was keen to have branches set up in other States and asked Thistle to organize an inaugural meeting in New South Wales. In Newsletter No. 1 in May 1957 there appeared for the first time the name "Society for Growing Australian Plants" (SGAP). Meanwhile Thistle, increasingly occupied with David's failing health, asked Mr N. H. (Bill) Hicks to chair a meeting in Sydney to set up a New South Wales branch. The meeting was held on Thursday, July 25, 1957, at the Real Estate Institute Hall, in the basement, 30A Martin Place at 8.00 p.m. There were seventeen people present, excluding Thistle, who would not leave David at this time. A second meeting was held in November 1957 just to check progress. The third meeting, on January 31, 1958, was the first Annual general Meeting, and Thistle was present to be elected as Vice-President with Bill Hicks as President.[28]

Following the death of her husband, David Stead, Thistle became even more involved than before in conservation matters and committed to the move by the Wild Life Preservation Society to find a fitting memorial for the man who had given direction to her life for almost forty years. As family members said, she became difficult to live with and preoccupied with matters that took her away from the house at Watson's Bay. It seems appropriate to record for the interested reader a summary of a number of issues with which she was involved during the course of one year, namely, 1958.

The Diary of Nature's Advocate for 1958

In 1958 Thistle Stead was still in full employment as a lecturer at Sydney Teachers' College. She was Honorary Secretary of the Wild Life Preservation Society of Australia, and Editor of its publication, *Australian Wild Life*. Information on Thistle's activities for 1958 has been taken from that publication.

Item 1 — For many years the Wild Life Preservation Society had been working actively for the elimination of grazing from the alpine areas in Australia. Following investigations by a committee of the Academy of Science (The Joint Advisory Scientific Committee), State Cabinet decided to allow grazing on the snow leases by permissive occupancy until June 1958, after which no grazing was to be permitted above 4,500 feet. However, early in the year, some press publicity was given to notice of a resolution by Mr J. Seiffert, M.L.A., in the Labor Caucus to restore grazing leases in snow areas in the Kosciuszko State Park by alteration to legislation to make the Catchment Area Protection Board a recommending body with no right to veto, in matters of land administration.

A memorandum was drawn up by Dr R. Garrick and Mr A. Costin of the C.S.I.R.O., and Thistle attended a meeting in the office of the Director of the Australian Museum to discuss the contents of this memorandum, which argued for the necessity of an early declaration and definition of a Primitive Area within the Kosciuszko State Park. The meeting was attended by representatives of the major Conservation Departments and voluntary organizations of Sydney and Canberra, the contents of the memorandum being discussed and agreed upon.

Following the meeting, Thistle as Secretary of the Society, sent a letter to the Acting Premier of New South Wales, and another one to the Editor of the *Sydney Morning Herald* (this letter was not published).

Item 2 — Following on the advice of Mr Roy Bennett, Council member, that there was danger of alienation of brush land bordering the Minnamurra Falls Reserve, New South Wales, endangering the Reserve itself, Thistle visited the Reserve and interviewed the Ranger, Mr Howard Judd . . . The Reserve itself is kept in excellent condition by the Ranger. I have never seen a Reserve better cared for, Thistle said. She wrote to the Kiama Council which controls the Reserve and congratulated it on the way it is maintained.

Item 3 — Thistle and the President, Mr W. H. Childs, had several press interviews about the proposal to send koalas to zoos in the United States. Personal representations were made to the Chief Secretary, Mr C. A. Kelly, and the Minister for Trade and Customs, Canberra, Senator N. H. D. Henty, urging that permission be withheld. Mr Henty was reported to have said that he would not issue a permit for their export, but, immediately following the State elections, Mr Kelly was reported as saying that he would not object to the export of six koalas to San Francisco subject to certain conditions.

Item 4 — About the middle of 1958 the Society received a request from the Horticultural Sub-committee of the Sydney Committee (which organizes the Waratah Festival) to sponsor an exhibition of garden-grown wildflowers during the Festival. An exhibit was duly staged, and Thistle contributed a section in a four-page brochure for distribution to visitors, her section being "Protecting Our Wildflowers".

Item 5 — A Conservation Conference was held in Sydney in 1958, the WLPSA being represented by Messrs R. F. Bennett and C. R. Ockenden. The preliminary work carried out by the Organizing Committee (of which Thistle was a member) in arranging the motions to be forwarded in some logical sequence, and arranging for a limit of time for each speaker, resulted in the Conference being conducted efficiently.

Item 6 — Thistle was the Society's nomination for 1958 for the award of the Australian Natural History Medallion. (She won it in 1963.)

Item 7 — Thistle was strongly committed to the progress of the David George Stead Memorial Fund, opened in 1958 after Stead's death in 1957. A committee set up by the Society (Thistle was on the committee) favoured the publication of a volume showing the nature and distribution of National Parks and Reserves in the State of New South Wales, a project which David Stead had always hoped to see achieved during his lifetime.

Item 8 — Thistle spent an afternoon with a new group of conservationists in the vicinity of Fullers Bridge on the Lane Cove River, exploring a reserve which they hoped to get authority to supervise. She reported that it was a beautiful sandstone area, with an excellent Christmas Bush swamp included, and expressed the opinion that it should prove of great value as a well-cared-for reserve.

Item 9 — Thistle visited Dubbo during the year to support the work of the National Arboretum Society (Dubbo). She expressed admiration for the work being done there: three hundred plants put out this year, to replace to some extent the plants lost in the drought of 1957; a herbarium of local plants (within an area of 30 miles around Dubbo) contained 400 labelled specimens to date.

Item 10 — The Council of the Society (WLPSA) showed much interest in the development of the Lake Parramatta Reserve, an area of approximately 181 acres at North Parramatta, some of which had been converted into picnic and playing areas but most still existing as virgin bush. Mr J. G. McKern (one of the Society's Vice-Presidents) had drawn the attention of the Society Council to the need to safeguard the area. Thistle, as Secretary, wrote to the Mayor of Parramatta on May 3, 1958, pointing out concerns about destruction in the Reserve and making suggestions about its management. As a result of this letter, Mr McKern and Thistle were invited to attend the next meeting of the Parramatta Ward Committee of the Council to speak personally on the matter. This they did, receiving a most courteous hearing. They were invited to submit a written report to the Parramatta Council, and drew up what would be considered today as an Environmental Impact and Assessment Statement. The proposals put forward by the Society in this report were discussed by the Parramatta Council and the Mayor said that "they were all in favour of them and intended to preserve the area as was suggested". The matter received favourable mention on the air (2FC) and in the *Parramatta Advertiser* of 21.10.58.

Item 11 — The Wild Life Preservation Society was working in co-operation with the Gosford Fauna and Flora Protection Society in an attempt to get a reasonably large area set aside at Bateau Bay as a Reserve . . . In November 1958, Thistle accompanied a party from the Gosford District Fauna and Flora Protection Society and the Bateau Bay Progress Association which took the Minister for Lands and the Deputy Surveyor General to inspect the land with a view to reservation. Following this a letter was received from the Minister indicating that certain small sections were to be reserved for Public Recreation and the Preservation of Native Flora.

The above items, not a full record of Thistle's activities by any means, indicate to what extent she was involved in matters of conservation — the diversity of commitments in terms of particular issues, the range of localities in which those issues were addressed, the secretarial work added to the meetings, discussions and addresses to groups, the obvious respect held by various bodies for her expertise and advice — all this speaks for an energetic, committed, and capable woman of action.

Thistle used to drive to Teachers' College each day. David used to stand in the street each morning like a traffic policeman, directing Thistle as she backed out of the driveway into the street and stopping any traffic in the rather narrow street whilst the manoeuvre took place. David's grandson, John, recalls that "Thistle was not ever a brilliant driver and had many minor accidents over the years (and a couple of major ones)." But John added that the young ones were rather proud of Thistle's driving because not many women drove at that time and it was quite something to be dropped off somewhere and being able to say that "my grandmother drove me in her car". In spite of this, the young Stead children were aware of her erratic driving habits which for young children had a daringly reckless air of bravado about them but as adults caused considerable consternation. In later years John refused to be in a car if Thistle was driving and always insisted on taking the wheel.[29]

In the years 1957–59, Thistle's niece Carol, and David's grand-daughter Dorothy, were attending Sydney University. Dorothy, who lived at Boongarre with her parents, along with Thistle in her part of the house, had the privilege of driving Thistle to and from College each day to practise her driving. Occasionally Carol and Dorothy would go down to the Teachers' College to have lunch with Thistle in the lunch room.

It was also about this time that Thistle refused the offer of an MBE, both she and David being strongly opposed to imperial honours.

From 1957, the Society for Growing Australian Plants in New South Wales was going from strength to strength in terms of membership. Thistle had been the Senior Vice-President of the Society since its inaugural AGM in January 1958. Mrs Carmen Coleman was Honorary Treasurer in the early days, and also acted as Editor of the newsletter for some time. Thistle was very active in the Society in the early days — in a circular for July 1, 1959 she is named as one of three delegates involved in arrangements for the Waratah Floral Festival to be held in October in the Town Hall, Sydney. At the General Meeting of the New South Wales Division of SGAP held on Friday, July 17, 1959 at the Sydney Technical College Theatrette, Harris Street, Ultimo, Thistle was the speaker, "Some Native Flower Gardens in Australia".

In August 1959 the SGAP Country Letter carried a short paragraph about Thistle. "The Society is in an enviable position indeed, in having a Vice-President of such high calibre. As most of you know, Honorary Life Membership was conferred on Mrs Stead at the inception of the Society in recognition of her work for the preservation and perpetuation of Australian flora and fauna. Mrs Stead is also the indefatigable Honorary Secretary of the Wild Life Preservation Society of Australia.

In December 1959, the New South Wales Society published its first issue of *Australian Plants*, a journal which was separate from the newsletter. Bill Payne was the editor of *Australian Plants*, and from the first day he accepted this position Thistle was outspoken in her opposition to him, claiming he didn't have the experience or the ability. Some years later she was to concede that Bill had produced a first-rate publication.

By early 1960 there was dissatisfaction within the New South Wales Division of the Society. Rupert Hill had taken over from Bill Hicks as President and some members thought his leadership was not strong enough. But it was Thistle who was most concerned about the leadership of the Society and the direction it was taking. She believed it was time she took the President's position and placed her supporters in key executive positions. Therefore, to ensure a successful outcome for her goal at the forthcoming AGM in March 1960, she organized a meeting of her supporters at the YWCA premises in Liverpool Street Sydney, about a week before. One of those who attended that meeting was John English, a newcomer to Society meetings and today still wondering why Thistle extended him an invitation. He was rather taken aback by the proposed line of action, thinking it was perhaps just a power play and unaware of any hidden agenda. The next morning going to work on the train, he shared his misgivings with Society member Max Hewett, who promptly raised the alarm with other members, including Henry Boyd.[30]

Thistle had drawn in around her a coterie of women, one or two of whom were intemperate in their outspoken contributions to discussions, and this did not endear Thistle, or her supporters, to the male members. Henry Boyd was particularly antagonistic and supported the drive to raise up an opposing faction. There was a general fear, which may have been irrational, that Thistle would move the Society in a direction not acceptable to the bulk of the membership. General gossip claimed that Thistle was concerned about competition for her books from Bill Payne's journal; the membership was highly satisfied with the journal and feared its demise under Thistle's leadership. The Society was going well since its start in 1957 — "in those days it was very much a grower's society, like a garden club, not conservationist. Maybe Thistle wanted to swing the Society towards conservation," was one comment from a member. "Maybe she was stirred to take action because of Rupert Hill's lack of strength," said another. Yet another member from those early days expressed the opinion that Thistle probably regarded the SGAP as hers, and had a proprietoral interest in setting it in the direction she thought was best. Whatever her motives, however, today we remain ignorant of them.

The opposing faction "pulled someone out of the woodwork, an orchid man who didn't have much knowledge about natives". This was Charles Taylor, who was to be the Society's President for the next nine years, "and he did a good job".

The third Annual General Meeting of the New South Wales Region of the SGAP was held on March 18, 1960, at Sydney

Technical College Theatrette, Harris Street, Ultimo. Almost 100 people attended, "the meeting stacked with people I'd never seen before," said one of those who were present. Dr Bartholomew chaired the meeting and supervised the voting. There were two nominations for President, Thistle Stead and Charles Taylor. Dr Batholomew would not reveal the detailed results of the secret ballot, but announced that Charles Taylor was the elected President, with Vice-Presidents to be Mrs T. Y. Stead and Mr R. L. Hill. Then followed the election of other office bearers: Secretary, Mrs A. N. Croucher; Treasurer, Mrs C. Coleman; Journal Editor, Mr W. Payne. It was a stormy meeting, Charles Boyd recalls,[31] with intense lobbying going on, people calling each other names. "I was ashamed at the bickering going on," he added.

Thistle Stead may have suffered a setback but she did not retire completely from the scene, although several of the early members recall that she did not attend many meetings after that AGM of March 1960. In the spring of 1960 she was one of three in a panel of judges who had the task of visiting the thirteen gardens which were nominated by the various districts for the final selection of winners in the garden competition. For the Newsletter in November 1960, Thistle wrote over two foolscap pages of comments on various gardens that were visited. She wrote, "All three of us were impressed, not only with the quality of the gardens, but also with the deep satisfaction which the competitors obviously took in them and which made our task an altogether enjoyable one".

In November 1960 Thistle was the guest speaker for the North Shore Group of the SGAP, speaking on "The Identification of Native Plants". The address was illustrated profusely with excellent colour slides, and questions were asked by members as the talk progressed. Thistle showed the slides arranged in families and discussed seven of the most common families in detail. "All in all, a most informative evening," the Newsletter reported. Max Hewett believes that the undercurrent of dissatisfaction with Thistle Stead in some quarters of the SGAP did not refute the fact that Thistle Y. Harris remained a certain recognized authority on the Australian flora, at least up until her later years.[32]

In November 1960, the Council of the Municipality of Woollahra had been asked to consider a proposal to establish a Women's Bowling Centre at Gap Park, Watson's Bay. The matter was referred to the Works Committee, and in December when that Committee presented its monthly report, Alderman Belle Miller, supported by Alderman Hugh Foster, asked for an amendment, "That the Council endorse-in-principle the establishment of a Women's Bowling Centre within Gap Park." After debate on the amendment, it was withdrawn by the mover and seconder. The Works Committee had asked that consideration be deferred pending the submission of plans sufficiently detailed as to enable a complete appreciation of the proposal to be made, and that the Engineer submit a report on the provision of access to Gap Park generally, via the disused tramway area.[33]

Alderman Miller submitted at length the results of investigations into the proposed Women's Bowling Centre in January 1961, setting out details of the cost of greens and clubhouse, and the method of financing the proposal. The Minister for Lands had indicated that he was awaiting Council's concurrence in the matter. It was resolved by Council at this January meeting that subject to the concurrence of the Minister for Lands, the Council agreed to the establishment of a Women's Bowling Centre in Gap Park near the former railway loop, to be known as the Wentworth Women's Bowling Club.

On January 31, 1961, Thistle wrote to Woollahra Council expressing concern that one of the small remaining open spaces at Gap Park could be the site of proposed bowling greens, and stating that it should be regarded as belonging to the whole of the people and not to a selected few. Mrs Marian Hill, another local resident, also wrote on 31 January to state that she considered the proposed bowling club at Gap Park to be a retrograde step, the area being one of the few pieces of free natural bushland left in the Municipality. Another lady wrote to

Council to suggest that the persons who would use the Bowling Club would be more profitably engaged in charitable activities. Mrs Hill wrote again in February to advise Council that a committee had been formed to protest against the establishment of the Women's Bowling Club, that a petition had been circulated in the immediate vicinity of the proposed club and approximately 300 signatures had been obtained expressing disapproval of the scheme. The petition was to be forwarded to the Minister for Lands. Thistle Stead was the spokesperson for this committee, and she was invited to address Council at its next meeting.

Consequently at an Ordinary Meeting of Woollahra Council on March 27, 1961, Thistle made representations on behalf of a section of Vaucluse residents in opposition to the proposed alienation of about two acres of the ten acres of Gap Park for the establishment of a women's bowling club. She stated that the residents she represented, who had formed themselves into a Committee, did not wish to see any further encroachments into public reserves for any minority use, particularly this section of natural bushland on which grew a wide variety of interesting and rare plants. Thistle answered a number of questions asked by aldermen, following which the Mayor (Alderman A. C. Murchison) stated that the Council had already forwarded the proposal to the Minister for Lands for his determination and there would be no further action at present. The following day, Mrs Marian Hill wrote to Council on behalf of the Committee for the Preservation of Gap Park to express thanks for allowing their spokesman, Mrs Stead, to address Council and for the Mayor's courtesy and impartial chairmanship during the address. Mrs Hill added that the Committee was not connected with any political party.

It would appear that Thistle considered the support of the Wild Life Preservation Society should be sought in the matter of the Women's Bowling Centre at Gap Park. In the Supplement of *Australian Wild Life* for September 1961 reference is made to Thistle's advice to the Society regarding Woollhara Council's recommendation to the Minister for Lands. "The Society's intervention, along with that of other interested people, seems to have the recommendation held in suspense for the time being," the notice in the Supplement reported. In *Australian Wild Life* for June 1962, the Society reported, "Members will recall reading in the Supplement of our efforts to prevent Gap Park being used for a women's bowling green. The Minister for Lands has recently informed us that he has refused permission for the park to be used in this way."

The correspondence to the Council from the Minister for Lands had stated:

> "A great number of objections to the proposal was received . . . and the matter was carefully investigated by the Metropolitan District Surveyor. The Minister is satisfied that the merits of the proposal are not sufficient to outweigh the need to retain the whole of Gap Park for future development and unrestricted public use." (Council Minutes, 11.12.61, p. 8)

Considerable improvements were carried out at Gap Park in the mid 1960s. This work had cost $5,000 and Council, at the Council Engineer's request, voted a further $5,000 to carry out additional works including water supply, grassing, fencing and crossing construction.

In August 1967 Thistle wrote congratulating Council on its choice of shrubs recently planted in the Park. Evidence of her continuing interest in the reserve can be found in an article which appeared in the *Wentworth Courier*, 18.10.72, p. 14, reporting that a schematic plan of improvements for Gap Park had been prepared by Mrs Stead at Council's invitation, due to her interest in the area. Part of the plan was adopted by Council in October 1972.

Thistle retired from her position as lecturer at Sydney Teachers' College at the end of 1961, the official date of retirement being February 25, 1962. Dr Turner, the Principal, wrote a charming letter to her, with an invitation to attend a function early in Term 2, 1962, to say an official goodbye and

Thank You. His letter contains a most fitting appraisal of the niche she had occupied in that establishment for twenty-four years. "I should like you to know how much your colleagues have valued the contribution you made to the College during your long period of service with us. You brought to your College work the kind of vitality, wide knowledge and experience that inspires students; I know you have led many of them to become active workers in Natural Science and Biology. Your field of exploits with students and staff have yielded stories that are almost legendary; I feel they rightly place you among the more adventurous of our colleagues and certainly prove you one of the most interesting personalities the College has had.

"We on the staff enjoyed your company, understood and appreciated your forthrightness, and, in the past, followed excitedly from the sidelines some of the more spectacular clashes of opinion you had with Dr Currey and others in staff meetings. Since you left us some of the verve has gone from these meetings."[34] Dr C. H. Currey was Head of the History Department up to 1951.

Since David Stead's death in 1957, Thistle and other members of the Wild Life Preservation Society had been considering the matter of a fitting memorial. Carmen Coleman, a member of the Society, had been looking for years for a suitable site on which to set up a Conservation Centre, a personal dream which came close to fruition when she found land at Bargo in southern New South Wales in 1962. After discussion with Thistle, the two women decided that this could become an appropriate memorial to a man who had laboured all his life to promote conservation. The story of the Wirrimbirra Sanctuary, as it came to be called, deserves a chapter or more in its own right, and the reader is invited to follow the ups and downs of this ambitious project in the chapter that follows.

Angus and Robertson published a small book by Thistle in 1962, *Eastern Australian Wildflowers*. In selecting the species to be included, Thistle was careful to bring in some of the alpine plants in which she was becoming interested. The publication wasn't very successful, Thistle said later, mainly because the publishers made the mistake of not releasing it with a glossy jacket, and it wouldn't have been able to compete with the glossy hard-covered small publications of people like Frank Hurley.[35]

In June 1962, Thistle's mother, Nana Harris, was expected at Watson's Bay on one of her regular weekend visits. Nana Harris would visit each of her daughters in turn each Sunday, in spite of having suffered a few minor strokes in the previous twelve months. This Sunday Thistle was away on one of her trips, but Ivy still expected Nana to call. When she failed to arrive, Ivy became worried and went to Mosman to check, finding the old lady in bed — she had suffered a major stroke. Ivy had her transferred to Boongarre and looked after her until Thistle returned. For six months Nana Harris stayed in David's old room, cared for by Thistle and Ivy until she died in November 1962, at the age of eighty-three.[36]

The Wild Life Preservation Society recorded with regret the resignation of Thistle as Honorary Secretary to the Society due to the illness and death of her mother. The Society acknowledged that "Mrs Stead's tireless work as Secretary for a number of years has been an inspiration, and we are most grateful for her personal contribution to the cause of wild life preservation. Happily she is able to continue as Editor of the Society's journal and as a member of Council."[37]

In the Supplement to the Journal in March 1963, Thistle gave a "Message from the Retiring Secretary". She said, "As many of you know this has been no sudden decision on my part although it was precipitated at this stage by personal commitments. I have felt for a long time that the Society was tending too much to become a Stead monopoly. My husband guided the Society from its inception in 1909 almost uninterruptedly until 1945, either as President, Vice-President, Treasurer or Secretary, often holding more than one office at a time . . . Mrs F. R. Hooten has now generously offered her services to the Society . . . One of the things I most cherish from my years of Scretaryship is the personal association I have had with so many of you and with many real friends I have made. It has been a rewarding job for me and I feel sure Mrs Hooten will find it equally so."

Spring of 1963 found Thistle in Western Australia, concerned that insufficient safeguards had been made to ensure the survival of its unique wildflowers. Quite recently, she reported, an attempt to amend the Native Flora Protection Act to ensure a greater measure of protection was defeated, but stricter regulations governing the picking of wildflowers under the existing Act were promulgated. Among the conservationists these new regulations had caused some concern because under the regulations, no-one was permitted to have in his possession at any time any part of any protected wildflower unless taken from private land with permission. This, according to the conservationists, was an unreal approach. Thistle referred to a statement by Vincent Serventy, then Supervisor of Nature Study in the Schools, who claimed that it was essential for teachers and children to be given some liberty to pick in order that they may teach and learn about the State's wildflowers. They would, however, like to see a licensing system introduced to restrict the sale of wildflowers. Thistle also talked with many teachers and was impressed with their appreciation and their teaching of conservation ideas.[38]

Early in 1964 Thistle received the news that she had been awarded the Natural History Medallion for 1963. This prized award of the Field Naturalists Club of Victoria was conferred annually upon distinguished contributors to natural history or science. Thistle had been nominated by the Wild Life Preservation Society of Australia, and an issue of its journal, *Australian Wild Life*, had paid tribute to its long-time member. "Mrs Stead's career in education and conservation is well known to the members of this Society of which she has been President (1949–50, 1951–52), Vice-President, Honorary Secretary, Honorary Editor and Councillor. Mrs Stead's published books under the name of Thistle Y. Harris are standard works of reference throughout Australia and have done much to promote a popular interest in wildlife as well as to extend academic knowledge."[39]

Thistle Stead was 62 years old. She was about to embark on another consuming passion, the memorial to the only man to whom she had given her total commitment. And she was still to produce four more books whose research was to be yet another consuming undertaking. But it seems appropriate at this stage in our story to go with Thistle for a summer sojourn to Kosciuszko, the high country which she frequented every summer for decades. There she immersed herself in its meadows of flowers, as well as its ravages by grazing and engineering works, at the same time making active efforts to preserve some measure of integrity for its more vulnerable areas.

Chapter 5 Notes

1. Unpublished notes in T. Stead's papers.
2. D. Bull, pers. comm. December 1996.
3. P. Macdonald, pers. comm. 30.11.96.
4. M. Reynolds, pers. comm. 2.7.97.
5. ibid.
6. I. (Robson) Cohen, pers. comm. 7.7.97.
7. ibid.
8. ibid.
9. K. Wall, pers. comm. 25.6.97.
10. ibid.
11. A. Strom, pers. comm. 2.1.97.
12. J. Wylie, pers. comm. 28.6.96.
13. B. Morley, pers. comm. 30.10.96.
14. L. Gilbert, *The Royal Botanic Gardens, Sydney*, Acknowledgements.
15. Transcript of L. A. Gilbert's response to the launch of his book at the Botanic Gardens, 1986.
16. K. Wall, pers. comm. 13.6.96.

17. J. Steanes, pers. comm. 16.6.97.
18. A. Berglund, pers. comm. 11.12.96.
19. L. A. S. Johnson, pers. comm. 23.9.96.
20. A. Clark, pers. comm. 23.1.96.
21. M. Reynolds, pers. comm. 2.7.97.
22. ibid.
23. ibid.
24. V. Serventy, pers. comm. 19.10.96.
25. "Selected Letters. Christina Stead", p. 163.
26. D. Basili, pers. comm. 26.11.96.
27. F. D. Hockings, *A History of the Society for Growing Australian Plants*, SGAP Journal, S.A. Region, Vol. 7, No. 9, 1977.
28. R. Doig, pers. comm. 11.9.96.
29. J. Stead, pers. comm. 14.7.97.
30. J. English, pers. comm. 15.12.96 and M. Hewett, 4.12.96.
31. C. Boyd, pers. comm. 18.12.96.
32. M. Hewett, pers. comm. 4.12.96.
33. This information on Gap Park, and that which follows was obtained from Council Minutes at Woollhara Council Library by Dorothy Basili and Jane Britten.
34. ML MSS 5715 Box 5.
35. de Berg Tape, Side 1, p. 12.
36. D. Basili and C. Hinder, pers. comm. May 1997.
37. Supplement to *Australian Wild Life*, December 1962.
38. Supplement to *Australian Wild Life*, December 1963.
39. *Australian Wild Life*, Vol. 4, No. 5, June 1965, p. 34.

CHAPTER 6

Kosciusko [1]

"I went to the woods because I wished to live deliberately, to front only the essential facts of life, and to see if I could not learn what it had to teach, and not, when I came to die, discover that I had not lived."

Henry Thoreau, 1817–1862

Thoreau had his Walden. Thistle had Kosciusko.

HENRY David Thoreau was a nineteenth century observer of nature who set up a hut by the brink of the Walden Pond in his native Massachusetts (USA), living there alone for two years to observe the natural world around him. He later described his experience in his book, *Walden*. It was said of Thoreau that although he was a true and passionate observer of nature, the poet in him was stronger than the naturalist, and he was for ever looking through the beauty of the earth in search of something behind and beyond. In Thistle, who was also a true and passionate observer of nature, the naturalist, the pragmatist, was stronger than the poet, and for her the beauty lay in nature itself and she had no need to search for something behind and beyond. Yet the only piece of truly creative writing found in her papers was an unpublished story undoubtedly inspired by her love of the snow country, "The Snow Princess".

Thistle was first and foremost a botanist, and it was her love of the alpine flora which took her every summer to the high country of New South Wales, occasionally to Victoria, and later to Tasmania. But she was a sociable creature and did not frequent the old settler's huts alone, and like Thoreau she could find poetry when expressing her delight in the flora of the alpine regions.

> " . . . on the very roof of Australia are to be found species of wild flowers unknown elsewhere, exquisite in their delicate beauty yet sturdy enough to withstand the snow-laden blizzards of winter and the hot drying winds of summer. In a good season they cover the moors in a blaze of colour, often brought to fulfilment by a few days of summer temperatures after a prolonged period of snow and sleet. From neat compact bushes, from twisted branches scrambling over rocks, from the sides of rippling streams along the line of the melting snowdrifts and from the heart of the peat bogs a riot of colourful blossoms appear as by a miracle, while the air is laden with the delicate perfume which so many of them possess."[2]

Since her first visit to Kosciusko as a university student, Thistle had been both fascinated and concerned about the fate of the relatively small alpine areas in Australia. When 1.3 million acres of Crown land were set aside as the Kosciusko State Park in 1944, under the management of a Trust, hopes were raised that this would result in effective protection for the unrivalled alpine flora and its concomitant fauna.

It was the uncontrolled grazing of sheep and cattle over a period in excess of 100 years that gave rise to expressions of concern to the Government that ultimately led to the establishment of Kosciusko State Park. The practice of grazing in the high country commenced some 160 years ago and it was not until the 1890s that some measure of control was introduced when the area was subdivided into snow leases and permissive occupancies. But grazing continued with little supervision until the 1930s, when Mr E. S. Clayton (then an officer of the Department of Agriculture) became concerned that soil was being blown and washed away and land eroded because of bad usage.[3] In 1942 Clayton approached the New South Wales Premier, Mr W. J. McKell (later Sir William McKell) and asked him to accompany him in a tour of inspection of the region.

This the Premier agreed to do, and his dismay and concern at what he saw led to the Kosciusko State Park Act in 1944. The year 1949 saw the passage of legislation to proceed with the Snowy Mountains Hydro-Electric Scheme, and this brought more problems for the Kosciusko State Park Trust. Grazing above 4,500 feet ceased as from May 1958, but continued in other parts of the Park until the passing of the National Parks and Wildlife Act in 1967. In 1969 a decision was made to phase out all grazing in the Park.[4]

The Wild Life Preservation Society of Australia took a keen interest in the conservation of the alpine areas, no doubt encouraged by Thistle and her life-long devotion to the snow country. David Stead, in his capacity as Acting Honorary Secretary and Honorary Editor of the Society, wrote an article for the journal *Australian Wild Life*, in June 1946, saying, "the recent establishment of the Kosciusko State Park, by the New South Wales Government, is far-and-away the most important move in wild life and native flora conservation and protection that has ever been made in Australia . . . [however] it is not at all clear from the wording of the Act of Parliament itself (Act No. 14, 1944) that this fauna and flora conservation is to be the first consideration in the future administration of the Park, whatever may be the intentions of the Kosciusko State Park Trust at the outset."

"From the moment the Kosciusko State Park Act was passed," Stead continued, "it has been made abundantly clear in Press comments and the statements of several parliamentary representatives and some Ministers of the Crown — not to mention the sports interests — that this Park must be 'developed' as one vast 'playground'." The article went on to say that, as a conservation Society, the WLPSA was not against the use of parts of the Park for alpine or other sports. But Stead pointed out that under the Act, the Park Trust was limited, in its disposition of the lands under its charge, to a total not exceeding one tenth of the total as a "primitive" area. David Stead had written to the Park Trust on March 6, 1945 urging it to make a general survey of the area before any decision was made regarding the provision of Wild Life and Flora Sanctuaries within the Park. He emphasized that from his personal knowledge of the alpine areas it would be quite impossible to get one self-contained sanctuary to embrace the faunal and floral features; he offered the opinion that there would need to be at least half-a-dozen separate major sanctuaries, with a number of others of smaller size preserved for some special floral features . . . The suggestion for one Primitive Area should not be entertained for a moment.[5]

In the early 1950s Thistle travelled to Kosciusko for her annual study of the flora, but also concerned to look at damage being done in the area by engineering works. For three successive years she was accompanied by a number of friends, which included her student Connie Burns, a member of the mature age intake to Sydney Teachers' College in September 1950. The first of these three trips was in January 1951, when they stayed at the Chalet at Charlotte's Pass, Thistle, Connie, and Mary Duckworth. In January 1952, three of them accompanied Thistle — Connie, Mary, and Jean Fountain, a teacher. The main purpose of this particular trip was a meeting with a certain Dr Mueller who drove them around the Snowy — "bad roads and another terrible driver". Dr Max Mueller was an environmental officer who started work with the Snowy Mountains Authority in December 1951. He had a European reputation for knowledge in the field of soil conservation with qualifications in engineering and agriculture. He was also a first-class taxonomic botanist, and lost no time in becoming familiar with the flora of the area, a study in which Thistle was to be of assistance. Dr Mueller's two main aims were to (a) carry out construction work in such a way as to minimize damage, including erosion; (b) to make immediate replacement or compensation where damage has been done, and natural cover removed, and thus avoid the results of erosion as far as possible.[6]

A large group of twelve made up the party in 1953. This visit followed an invitation from the Snowy Mountains Authority, tendered through Dr Max Mueller, to visit the area with which the Authority was concerned. Thistle and Kath Elger travelled in one car, and Frances Hill with Connie Burns in another car, these two cars meeting at the Eucumbene crossing. They arrived at dusk and saw two platypuses. Later they met Pat Bourke and his wife and child, two students and their brother, and Shirley, another student, and her son. They stayed in cabins left by the Snowy Mountains Authority, close to the river, and there were flowers everywhere, Connie recalls. Dr Mueller personally conducted the group to various sections of the Authority's project.[7]

It was Nana Harris, Thistle's mother, and Wendy, the younger daughter of Thistle's sister, Joan, who went with Thistle to Kosciusko in January 1954. They stayed in a cabin at Talbingo on the Tumut River, about 30 km from Tumut on the north-western boundary of the park. Thistle would go off botanising all day, leaving Nana and Wendy in the cabin area.

During her visit to Kosciusko in February 1955, Thistle had several discussions with Dr Mueller and Mr Van Lint of the Scientific Services Division of the Snowy Mountains Authority regarding reclamation work in the area. On her return to Sydney she sent a letter to the Commissioner, Mr Hudson (later Sir William Hudson), congratulating the Authority on the roadside reclamation work being undertaken in many sections. But in the same letter she drew attention to signs of the use of bulldozers on the Blue Lake track — on the descent from Charlotte's Pass to the Snowy River — and along the track on the ridge above Seaman's Hut, causing serious erosion, particularly on the steep descent from Charlotte's Pass. Vigilance in the interests of conservation in the area was asked for, she said. To this letter Mr Hudson replied, stating, among other things, that the tractors had been immediately withdrawn and the ridge above Seaman's Hut recovered with grass two to three days later. "You may rest assured." Mr Hudson said, "that the Authority is anxious to avoid permanently marring the beauties of the area".[8]

In spite of assurances from the Snowy Mountains Authority, the Wild Life Preservation Society was still expressing concern in March 1956 about erosion in the snow areas accelerated by excessive grazing and engineering works. As Secretary, Thistle reported in the journal, "Many interests are now concerned in protecting the area and your Society continues to press for the withdrawal of all grazing leases at the higher altitudes, for the greatest possible care to reduce damage due to engineering projects to a minimum, and for reclamation work to be carried out on areas already damaged."[9]

In 1958, following her January visit, Thistle wrote to the Park Trust to congratulate it on the excellent fireplaces and fire notices erected along the newly-made Geehi-Jindabyne Road (Alpine Way). However, she was still expressing concern about the continued grazing.

In April 1959 the first professional park manager, Mr N. C. Gare, was appointed by the Park Trust. Neville Gare had previously been an officer of the Forestry Commission and was highly recommended for his position as Park Superintendent, living on site at the Park Headquarters at Waste Point. He was to remain until January 1971, a period of almost twelve years, and a period of rapid development and conflicts between grazing and recreation, catchment protection and nature conservation. Thistle arrived in the Park for Christmas 1959 and accosted Neville Gare in his home at Waste Point. She asked where he came from, and when he said, Forestry Commission, she said, "Oh, you were one of those foresters," giving him a clear message that she had reservations about foresters in general.[10] Later Thistle came to appreciate and understand Neville's policies and philosophy; during all the years Neville had direct contact with Thistle there was much happening — he still had to deal with grazing, he had to try to pull the Snowy Mountains Authority into line, he had to establish some control over the ski resort development, and also work to arouse public awareness about the Park and its values. When Thistle arrived each January she would see developments such as the information centre being built, the rapid ski resort expansion, and rangers being employed and she would have mixed feelings about it all.

Thistle had a wilderness concept about Kosciusko, Neville Gare said, and she wanted it to stay that way — she appeared to have little sympathy with ski resort growth, and Kosciusko did have a monopoly on the snow area of Australia. Thistle seemed to have some difficulty understanding how Neville had to reconcile all the different issues involved. With the passing of the National Parks and Wildlife Act in 1967, Kosciusko became a National Park, and Neville then had to push on with the job of establishing the National Park, and developing an understanding of what a national park was meant to represent in terms of both usage and conservation — a very difficult equation.

When Thistle stayed in the Kosciusko Park she took up residence in a cabin, or even camped, but in Neville Gare's time it was mostly in Spencer's Cottage (known as the Spook House). In the years from 1955 to 1965 she was usually accompanied by Lois Sourry, whom she had met when Lois was President of the Gosford Fauna and Flora Protection Society (now called Gosford Wilderness Conservation Society). Andrew, Lois' husband, was in the party several times in that ten-year period — he remembers it as a rugged spot. James Spencer, who ran cattle in the area in the 19th century, built his stone cottage between 1865 and 1878; a brochure produced by Kosciusko State Park in the early 1960s stated that Spencer's cottage had been renovated and seemed destined to become a museum for the historic area. But this was not to be — Spencer's cottage was drowned about 1967 as Lake Jindabyne became part of the Snowy Mountains scheme. Thistle also stayed a few times at The Creel, a picturesque old establishment on the Thredbo River not far from Park headquarters. It was built in 1906–7 to accommodate workmen building the Hotel Kosciusko, but later became a fisherman's lodge, well-loved by many anglers. The Creel too has gone today, having been demolished in the sixties as the lake rose behind Jindabyne Dam.

At various times in the Christmas-January period Thistle would have other friends around her too, as well as Lois Sourry in those years 1955 to 1965. There was Alison Ashby, artist from South Australia, Kath Elger, Eva Symes, and Baldur Byles, who was a forester and a member of the Kosciusko Park Trust. Baldur had worked with Neville in producing a plan for the Park. Lois and Thistle would go out with their cameras, both keen photographers of the alpine flora.[11]

The Kosciusko State Park Act 1944 provided, *inter alia,* for the Trust to establish a "Primitive Area" of up to 10% of the Park's area, but nothing was done about this until a group from the Australian Academy of Science decided to stir things up. In 1958 a group of scientists and naturalists from the academy drew up a scheme for the establishment of a large primitive area in the summit region. Their proposal was in part critical of the Snowy Mountains Authority (SMA), and the Federal Minister for National Development, Senator W. H. Spooner, who was responsible for the SMA, publicly rejected it. Nothing daunted, the Academy pursued its proposal with the Kosciusko State Park Trust, which at a later date (1963) formally resolved to retain the Primitive Area as provided for in its Act.[12]

In 1962 the Wild Life Preservation Society noted with satisfaction the publication of a brochure by the Australian Academy of Science entitled: *The Future of the Kosciusko Summit Area: A Report on a Proposed Primitive Area in the Kosciusko State Park.* The Society's journal pointed out that the Society had been represented among the naturalists who together with the academy's scientists had made the submission in 1958. The Committee from the Australian Academy of Science included Professor John Turner (Melbourne University), Professor Bob Crocker (Sydney University), Dr W. R. Browne, Dr Max Day, and Alec Costin among others. The 1962 Report proposed an area of 70 square miles which included sites ranging in altitude from 1,500 feet to 7,300 feet and covered within this comparatively small area alpine and sub-alpine sections, a wider range of vegetation than could be found anywhere else in Australia. Alec Costin had said, "The alpine tract above the tree line provides the best example of high mountain vegetation in Australia, including some communities which are found nowhere else".[13]

Thistle's article in *Australian Wild Life* pointed out, "It is well known to those who have seen the area that it is not, strictly speaking, a primitive area since erosion through many years of grazing and firing has affected many parts of it. Since the exclusion of grazing animals considerable recovery has been shown and the area includes a splendid section of the typical vegetation. The Academy suggests walking tracks and the maintenance of a motor road to the summit. The Snowy Mountains Authority is not altogether in favour of the scheme although it supports in theory the setting aside of a primitive area."[14]

Since the original submission about the primitive area in 1958, Thistle had been actively lobbying on behalf of the Wildlife Preservation Society. One example is the following, as reported in the Society's journal for June 1960:

> Your Secretary [T. Y. Stead] has just returned from an interesting tour of alpine areas, including sections of the Kosciusko State Park. As a result of discussions with the botanists of the Snowy Mountains Authority, the newly-appointed Superintendent of the Park and other persons, the following suggestions have been sent to the Chairman of Trustees, Kosciusko State Park:
>
> > "Because of its increased importance and because of the increased and increasing tourist traffic in the area, may I urge the Trustees to give further consideration to the proposal put before them some time ago by Conservation Bodies (including the Academy of Science and this Society) for the setting aside of a NATURAL or PRIMITIVE AREA within the park proper, for the purpose of preserving a section of the unique flora of the alpine area. This spring the flowers were magnificent, but the number of tourists stopping at points of interest and trampling and picking flowers has to be seen to be believed. For immediate consideration I would suggest that the following steps be taken to safeguard the area: . . ." (here followed suggestions about erection of notices, removal of obsolete and decaying buildings, attention to the rubbish

tip at the Chalet, and the need for a publication on the natural history of the Park.)

This letter to the Trust concluded, "I would like to congratulate the Trust on the appointment of Mr Neville Gare as Park Superintendent. During my recent visit to Waste Point, Mr Gare gave us every consideration and his interest in the welfare of the Park was most evident".[15]

Another letter was sent by Thistle early in 1961 to Sir William Hudson, Commissioner, Snowy Mountains Hydro-Electric Authority:

"I have recently returned from a very pleasant visit to the Kosciusko State Park. It was pleasing to see that some of the old and dilapidated buildings have been demolished. There is still a remnant of the old Spencer's Creek hut that should be removed, as well as a fibro hut near David's Moraine opposite the Spencer's Creek weather Station.

"Many of the erosion control areas (experimental) are showing their effect, but there are many other places where these controls could be introduced.

"I am particularly concerned with the bad erosion in the valley below the Albina Track, between Albina hut and Seaman's hut, where the river has changed course many times, due no doubt to the continued traffic over many years (now stopped) to Albina hut. Could some erosion control measures be tried here?"

Sir William wrote that the remnant of the Spencer's Creek hut would be removed, although it was no longer the responsibility of the Snowy Mountains Authority, that the hut near David's Moraine was being repaired for use, and that the Soil Conservation Department was undertaking repair on the Albina Track.[16]

No doubt as a result of the lobbying by the Australian Academy of Science and other interested bodies, including the Wild Life Preservation Society, the Kosciusko State Park

Trust in January 1963 declared a Primitive Area of approximately 70 square miles in the vicinity of Mt Kosciusko.

The Sydney Morning Herald announced, "With the creation of a 70-square-mile primitive area around the Kosciusko summit, Australia has taken the first step along the 'wilderness trail' pioneered by America's national parks. In America, this primitive area would be called a wilderness park. Both terms are equally apt in describing the new reserve, which is the direct result of pressure by scientists and park authorities who want to retain part of the Kosciusko region as a wild area unspoilt by roads, motels, service stations and other forms of commercial exploitation."[17]

The aim of the scheme was to preserve a natural recreation and scientific area which could be used by skiers, fishermen, campers, hikers and tourists and yet not be destroyed in the process. "We have no intention of keeping people out of the area; in fact, the Trust is anxious to attract increasing numbers of visitors," said the Trust chairman, Mr Howard Stanley. "What we will do is prevent road and engineering works, building and other forms of commercial development which could interfere with the character of the area."[18]

But maintaining the integrity of the proposed primitive area would not be an easy task. Early in 1963 it was necessary for Neville Gare and Baldur Byles to engage in some intense lobbying which resulted in a considerable number of letters landing on the desk of the State Minister for Lands, Mr K. C. Compton. Until then the Minister had shown little interest in the problem but now decided to take action, meeting Sir William Hudson on Mt Kosciusko, surrounded by the Media. Leading trustees like Sir Garfield Barwick and Howard Stanley supported the Primitive Area idea, but Clayton, then retired and acting as a consultant to the Snowy Mountains Authority, opposed it if it compromised any SMA works.[19] Baldur Byles and Neville Gare had been supported in their lobbying through action taken by the Garden Club of Australia, Mrs Margaret Davis being President at the time.

In April 1963 Thistle wrote on behalf of the Wild Life Preservation Society to the Prime Minister, Mr R. G. Menzies:

> "I do know this alpine area very well and as a botanist regard it as one of the outstanding areas in the whole Commonwealth floristically, indeed unique in the world. The area under question must be preserved in its entirety if this beautiful and unique example of Australian flora is to remain. It is impossible to replace it or to compensate for damage done by large engineering works within the area.
>
> "Conservationists have worked hard for many years to have summer grazing stopped in these high areas and have been rewarded by remarkable recovery of damaged areas . . . After seeing the colossal damage which has occurred in many places within the high alpine areas such as Cabramurra, Guthega and many other places, result of major engineering works effected by the S.M.H.E.A., I cannot agree with Sir William Hudson that the proposed engineering works within the proposed primitive area will not cause considerable and irreparable damage."[20]

Mr Howard Stanley spoke to a meeting of the Wild Life Preservation Society on June 17, 1963, thanking the Society for its support to the Kosciusko State Park Trust in its attempts to have the newly declared wilderness area kept free of engineering works proposed by the Snowy Mountains Authority. He stated that both the Government of New South Wales and the Park Trust had received thousands of letters and telegrams, many from overseas.[21]

In the end the Primitive Area survived and the proposed Spencer's Creek dam (below the Chalet and outside the Primitive Area) was not built by the S.M.A. The aqueduct system and tunnel through Charlotte's Pass designed to feed upper Snowy River water into this dam were also abandoned. A planned surface aqueduct from Lady Northcote Canyon to Geehi Dam, which would have permanently defaced Watsons Crags and the western face of the Main Range, was partly replaced by a tunnel through the Watsons Crags spur.[22]

In December 1963 Thistle wrote in *Australian Wild Life:*

> "A semblance of peace has settled on the controversy outlined in earlier issues between the Kosciusko State Park Trust and the Snowy Mountains Hydro-Electric Authority over large-scale earth and engineering works proposed within the Kosciusko primitive area. At such a time we should be on our guard lest the very emphatic stand taken by conservationists be weakened by simple passage of time."[23]

In January 1964 a group of thirty-three members of the Garden Club of Australia visited Kosciusko, taken on a tour by Baldur Byles and Thistle Stead to see the flowers and to investigate the damage and the likely damage caused by engineering works. They were impressed with the beauty of the alpine flowers and left the Park determined to do all they could to save the Primitive Area from further depredations.[24]

But the Primitive Area survived and today, in the nineties, is incorporated in the current Kosciusko Plan of Management as part of Management Area A1 for management of outstanding natural alpine and subalpine resources. Much of it is of course subject to heavy use in summer by visitors on foot beyond Charlotte's Pass and from the Thredbo chairlift, so the area is still under great pressure.[25]

After many years of research, Thistle's book *Alpine Plants of Australia* was completed and published in 1970 by Angus and Robertson. She acknowledged that in preparing the book for publication she had been generously assisted by many people, including government officials and personal friends. For the examination and study of material in New South Wales she was particularly indebted to the one-time Kosciusko State Park Trust and the Park Superintendent, Mr Neville Gare, and Dr D. F. McMichael, Director of the National Parks and Wildlife Service. For help with field work in Tasmania she gave her thanks to the Scenery Preservation Board, especially to its then Superintendent, Mr J. B. Thwaites.

Personal friends who supplied material included Mr Alan M. Gray, Mr and Mrs Herbert J. King, Mrs Kay Geeves, Mrs Betty Gee, Miss Kathleen Elger, and Miss Essie Huxley. Mrs G. M. Holley and Mrs Susan Heins did valuable work with preliminary field sketches, while the line drawings for the final text were done by Linda Turner. Then there were the staff members of herbaria in Sydney, Melbourne, and South Australia, Dr R. Carolin of the Botany Department, University of Sydney, and staff of the University of Tasmania, Dr Winifred Curtis and Miss Heather Guilline.[26] Kay Geeves and Alan Gray often sent living material from Tasmania to Thistle, the plants packed in damp paper and sphagnum moss. Thistle also sent cuttings to Tasmania but her parcels often fell apart as she apparently had no idea of how to assemble a secure package.

"I found the alpine flora of both the mainland and Tasmania fascinating. This work had an added advantage for me in that it made me acquainted not only with the Tasmanian flora but with the Tasmanian people, whom I find quite delightful . . . People don't understand, I think, or don't know how tremendously beautiful the alpine plants can be. You get great fields of colour when you get a good season. This year I was at Kosciusko just before Christmas [i.e., December 1971] and there was practically nothing in the upper areas, but the montane and the sub-alpine slopes were magnificent . . . The beauty of visiting alpine areas with this strange alteration in the time of the best flowering is that every time you go you see a new plant in flower. This time I saw for the first time in full flower a plant that has interested me for over ten years, and that was the little alpine *Veronica* which was absolutely magnificent, I had no idea it could grow like this.[27]

In February 1968 the Minister for Lands in New South Wales, the Honorable T. L. Lewis, commissioned Dr Graham Edgar, a former Director-General of Agriculture, to investigate and submit recommendations into controlled grazing and longer term leases in Kosciusko National Park. As a result of Dr Edgar's recommendations in 1969 all grazing within the boundaries of the Park ceased.[28] However, illegal cattle were still in the high country in the early seventies. The National Parks and Wildlife Service brought Jack Noonan down from Kinchega to deal with the cattle problem in Kosciusko; he was a man with a great deal of experience in such matters.[29] In January 1973, just after extensive bushfires in the Kosciusko region, Thistle left Sydney with Elizabeth Chandler to stay near Currango in Kosciusko National Park. Currango Homestead, still standing in the Park at its northern end, was a pre-park mountain homestead occupied by Tom Taylor, a senior Lands Department Ranger in Kosciusko who was in charge of policing grazing for the Lands Department. After he retired, Tom and his wife were given approval to stay on, and they set up the homestead and its annexe as accommodation quarters.[30]

Thistle and Elizabeth arrived in the afternoon to meet other friends, Kath Groat and her friend, Ida, at an old hut formerly used by the cattlemen; these huts were being used by trout fishermen in the sixties and early seventies. The table was set for afternoon tea but the reception was very cool. It was very puzzling, this feeling of antagonism from the female proprietor of the huts. They learned subsequently that a plane had been flying over, and this woman, whose husband was a cattleman, knew that Thistle was a friend of Tom Taylor; she thought the plane was checking the men and their cattle, and that Thistle was responsible for this, but the plane had been checking the bushfires.[31]

For two weeks the hut was their base for breakfast and dinner, as every day they went out collecting and photographing plants. One day they went out in Thistle's station wagon and approached a causeway on a creek, a crossing which appeared far too narrow for Thistle's vehicle, but they crossed anyway. All day long Elizabeth was wondering, "How shall we get back?" Elizabeth got out of the car and gave directions, and over they went. But the next day Elizabeth said, "No, not that way again". Another day they were driving down a narrow road in the vicinity of Tantangara Dam, down a steep gorge and around a bend. Ida started screaming from the back seat (all four were in Thistle's car that day), she was not familiar with Thistle's driving and her total lack of awareness about the danger in any situation.[32]

In January or February of 1975 Thistle was off to Kosciusko again, this time with John Teulon and Susan Heins. Ben Swane's daughter Elizabeth, then nineteen or twenty years old, asked if she could be part of the group. Accommodation was at the Yeti Alpine Ski Club lodge in Perisher Valley. They travelled in one of the vehicles from Swane's Nursery and had a most rewarding experience because Thistle had keys to all the gates which enabled them to travel on roads not accessible to the general public. John remembers Rawson's Pass and the Lake Albina area, and remembers too watching Thistle take up plants by the roots as part of her collection for Wirrimbirra. Of course she had a permit to collect, but what were its limitations? Thistle kept all their eating utensils in a picnic basket. Before dinner one night John offered to take the rubbish down to the collection area, but he was feeling very nauseated, and wanted to go to the hotel to get a brandy to settle his stomach. Elizabeth asked to go too, and in the vehicle told John how nauseated she felt. He said, I'll get some brandy to deal with that. Back at the lodge they could not disguise their recent exploit, and Thistle was horrified, "A man of your age!" she said to John. But the next day Elizabeth said to John that she believed she had discovered why they had a stomach upset — all the plates and other utensils went back into the basket without a wash, just a quick wipe over with some paper.[33]

Thistle's tale of the Snow Princess is a simple story of magic and romance, a fairy tale for children. It was obviously a product of the enchantment she felt herself for the snow country, although it only evokes pictures of the landscape to a small degree. The language is descriptive yet simple — it was fitting for a tale for children — and it was a story without a moral, for Thistle was not a moralist. Two children, Elsa and her brother Derek, go with their mother to the snow country to enjoy a holiday, planned to help Elsa recover from an illness. On the mountain hillsides that summer they enjoyed wandering through the flowers and sitting beside the "rippling brooks or the blue, blue lakes".

When the snows came the children made a snow princess and decorated her with blue beads for eyes and golden curls cut from Elsa's own hair. Elsa, who had been blessed by the fairies at birth, kissed the figure of snow and said, "I wish you were alive".

That night the Snow Princess, now alive, travelled by dog sledge to the great white palace, the home of her father and mother, the Snow King and Snow Queen. But alas, a richly clad young nobleman who fell in love with the Princess, lost her as she turned into a pool of clear water, leaving only two bright blue beads and a mass of golden curls. Early next morning the nobleman travelled with his sledge to the cottage of Elsa and Derek and flung the beads and the tangle of golden curls away from him. Elsa ran out that morning to look for her Princess, only to find she had gone, leaving behind the beads and the golden curls. "If only she and Derek could have known all the strange happenings of that night. But they were never to know. A blue, blue lake, frozen hard in winter and too deep to be crossed in summer kept the secret of the palace of the Snow King and Snow Queen from them forever."[34]

Many years after the days of botanising and lobbying Thistle wrote to Neville Gare, then Deputy Director of the Australian National Parks and Wildlife Service. She said, "It was nice to have a letter from you and I am grateful for the publications you sent which will be helpful. I am glad you speak so highly of Kosciusko and the state of its flora at present. Last time I was there I stayed at Perisher and while there were still plenty of lovely places to walk in I felt the old friendly atmosphere had given place to a new sophistication and a division of the population into two distinct categories. I don't really want to see it again."[35] It was 1985 and Thistle was 83 years old. It was the end of an era.

Chapter 6 Notes

1. The spelling of "Kosciusko" has now been changed to "Kosciuszko" but for the purpose of this account of Thistle's relationship with the area I shall use the spelling that was common usage through the time she knew it.

2. *Alpine Plants of Australia*, 1970, Introduction.
3. Information from H. Stanley, A *History of the Establishment of Kosciusko National Park*, NPSW, n.d.
4. ibid.
5. *Australian Wild Life*, June 1946, p. 12–13.
6. T. Y. Harris, "Our Alpine Heritage", *Australian Wild Life*, 1954, p. 30.
7. C. Burns, pers. comm. 20.8.96 and 9.5.97.
8. *Australian Wild Life*, March 1955. The Commissioner, Mr Hudson, was created a Knight Commander of the British Empire in 1955.
9. *Australian Wild Life*, March 1956, p. 5.
10. N. Gare, pers. comm. 17.3.97.
11. A. Sourry, pers. comm. 16.11.96.
12. *S.M.H.*, January 10, 1963.
13. *Australian Wild Life*, June 1962, p. 9.
14. ibid.
15. *Australian Wild Life*, June 1960, p. 4–5.
16. ibid.
17. *S.M.H.*, January 10, 1963.
18. ibid.
19. N. Gare, pers. comm. 4.7.97.
20. *Australian Wild Life*, December 1963, Supplement, p. 6.
21. ibid.
22. N. Gare, pers. comm. 4.7.97.
23. *Australian Wild Life*, December 1963, Supplement, p. 6.
24. *Australian Wild Life*, June 1964.
25. N. Gare, pers. comm. 4.7.97.
26. *Alpine Plants of Australia*, Acknowledgments.
27. de Berg Tape 574, Side 2, p. 1.
28. R. B. Good, *Kosciusko Heritage*, NPWS, 1992.

29. A. Fox, pers. comm. 20.3.97.
30. N. Gare, pers. comm. 17.3.97.
31. E. Chandler, pers. comm. 3.1.97.
32. ibid.
33. J. Teulon, pers. comm. 3.6.97.
34. Stead papers ML MSS 5715, 2/25.
35. Letter supplied by N. Gare, 17.3.97.

CHAPTER 7

Wirrimbirra — 1962–1975

"Mankind is part of the natural environment. We live in it. It sustains us both materially and spiritually. We need the peace and the serenity bushland provides. We cannot fulfil our potential without knowing it intimately."

T. Y. Stead

WIRRIMBIRRA Sanctuary at Bargo, south of Sydney in New South Wales, was established as a memorial to David George Stead, but it is also a memorial to Thistle Stead, who worked tirelessly in her later life to make the Sanctuary a worthy contribution to research and education on behalf of the Australian natural environment. It is also a living memorial to people like Carmen Coleman, who conceived the original idea, and to the Rangers who laboured there in drought and fire, and to the many volunteers, too many to be named, who over the years made the Sanctuary a place of beauty and utility.

In 1958, after David Stead's death in August 1957, The Wild Life Preservation Society established the David George Stead Memorial Fund, to be used, according to the response of subscribers, for one or more of the following purposes:

1. A scholarship or some annual assistance for work in fields that would benefit Nature Preservation or allied subjects.
2. An annual grant for some essay competition or a special piece of research in the fields indicated in (1) above.
3. An annual lecture by one of the Learned Societies.

4. A plaque to his memory in some reserve or in the rooms of some suitable Society.[1]

Circumstance decreed that the property now known as Wirrimbirra would be acquired in 1962 and with Stead family agreement, this became David Stead's memorial; finances from the David George Stead Memorial Fund were diverted to this project.

Carmen Coleman, a Sydney accountant and member of the Wild Life Preservation Society of Australia, discovered, after a long search, an area that she considered suitable for the establishment of a Conservation Centre where concerned environmentalists could meet, discuss, and implement activities to involve others in environmental conservation. She searched for several years on weekends before finding a suitable bushland area within easy reach of several highly populated centres. The site was Portion 32, Parish of Bargo, Shire of Wollondilly, 66 acres (26 hectares) on the Hume Highway between Tahmoor and Bargo.[2] She paid a deposit of £200 (which she had to borrow) and took out a mortgage, dated July 19, 1962, to Alfred John Harry of Epping.[3] Carmen was registered as the proprietor of the land, but she soon captured the interest of Thistle in the project, and Thistle paid the full price still owing, the mortgage being discharged in January 1963. The owners discussed ways of forming a company to administer the area and decided to assemble a group of people to propose a line of action. Thus, in 1963, under Carmen's guidance, the Articles of Association for the David G. Stead Memorial Wildlife Research Foundation of Australia (Incorporated) were drawn up and the name Wirrimbirra (dialect of a Goulburn Aboriginal tribe meaning "to preserve") was given to the area.[4] This Foundation became the registered proprietor of the land in August 1964.[5]

The principal aims and objects of the Company were:

- to initiate, promote and further research, scientific, technical or other, into all aspects of the conservation of wildlife of the Commonwealth.

- to maintain sanctuaries, reserves, and to carry on experimental work.
- to co-operate with organizations, universities and institutions in experiment or research in use of national parks, playgrounds and similar areas; in the proper management, cultivation and preservation of Australian indigenous fauna and flora.
- to publish information related to these aims and objects and to arrange lectures and carry on educational courses for the furtherance of these objects, and to establish fellowships and scholarships for work in conservation.
- to establish and maintain museums, arboretums and collections, related to its objects.[6]

This was a most ambitious set of objectives, and in the early days, Wirrimbirra was described as the Foundation's first project.

The first general meeting of the association was held on April 11, 1964 at "the project site of Wirrimbirra". The members of the first Board were N. H. Hicks (Chairman), Thistle Y. Stead (Deputy Chairman), H. J. Stanley, B. U. Byles, A. A. Strom, I. F. Wyatt, with C. R. Ockenden as Honorary Secretary and Robert L. Stead as Honorary Treasurer.[7] Ivor Wyatt, closely associated with the National Trust of Australia (New South Wales), of which he was a foundation member, and President for many years, was subjected to a great deal of pressure from Thistle to join the Board of the Stead Foundation. Ivor's mother, Mrs Annie F. Wyatt, O.B.E, who was responsible for the founding of the National Trust in Australia in 1945, had been a great friend of David Stead, and Ivor knew Thistle from early days in the Nature Conservation Council and other organizations. Thistle wanted Ivor's guidance and advice, she said, but was possibly also looking ahead; Ivor was appointed as the Trust's representative on the Board in 1964 and served until 1993, when he retired from the Board due to ill health. From 1979 to 1986 Ivor served as Vice-President, and from 1986 to 1993 held the position of President of the Stead Foundation.[8]

In 1964 the property was 66 acres, a 6 acre block facing the Hume Highway, and a 60 acre block on the other side of the main southern railway line. The objective of the Foundation for Wirrimbirra was a 200 acre block comprising the original 66 acres, 38 acres adjoining the block available for purchase, and about 100 acres in adjoining Crown leases, for which application had been made. Mr George W. Sample of Geo. H. Sample and Son Pty Ltd (Automotive Products) gave the £800 needed to buy the 38 acres, and Marie Byles, Solicitor, arranged for the transfer of title to the Foundation.[9] Thus the property came to include Portions 32 and 17 of the Parish of Bargo, traversed by the main southern railway, which cut off approximately 12 acres between it and the highway, the main area of nearly 90 acres being situated to the east of the railway line. The small section of 12 acres was established as "Wirrimbirra Gardens", where attractive trees and shrubs were planted, and the buildings erected. The entrance to the property on the Tahmoor side now had a set of wooden gates, the gift of Thistle's colleagues at Sydney Teachers' College on her retirement and a plaque to this effect was attached to the gates, long since removed.

Carmen Coleman worked out the first development plan for the area, including the planning and erection of the first building (the office), and the siting for the Ranger's cottage. She also worked out the first landscape masterplan, and the overall plan for the use of the area as an educational and research centre.[10] In a 6 acre block of Wirrimbirra Gardens, architect Milo Dunphy and landscape gardeners John Temple and Ken Harris spent a long day in mid 1964 mapping and measuring the area for the purpose of designing a landscape plan to modify the original one designed by Carmen Coleman.[11]

Thistle held concerns from the outset about the future security of this memorial to David Stead. Therefore, on the urging of the Foundation and especially the lobbying of Thistle Stead, The National Trust of Australia (New South Wales) agreed to accept Wirrimbirra Sanctuary, without payment, as one of its properties, the Trust President at the time being The Honorable Mr Justice J. McClemens. A notice dated November 7, 1965 was sent from the Foundation Board to all members convening an Extraordinary General Meeting on November 30 to discuss the transfer of the property. At the meeting 111 members were present, and the transfer authorized by the vote without any opposition. The circular of November 7 had emphasized that there was no question of The Trust wishing to take over the management of Wirrimbirra, "this will remain the prerogative of your Board and the members of the Foundation. Nor would the transfer of Wirrimbirra affect the independence of the David G. Stead Memorial Wildlife Research Foundation of Australia which will remain a separate corporate body registered under the Companies Act. The transfer to The Trust relates only to the freehold of the Wirrimbirra site for which your Board wishes to have guaranteed preservation in perpetuity."

From December 1965, Wirrimbirra therefore became the property of The National Trust and was leased to the Foundation for the purpose for which it was first acquired, the Sanctuary rent to be one pound per annum. Thistle had proposed this measure with the objective that the valuable Wirrimbirra site would remain inviolate for all time as a wildlife research establishment and flora and fauna reserve, in the permanent security and authority of the Trust's patronage.[12] With the approval of the National Trust President Ivor Wyatt had negotiated with the Trust on behalf of the Stead Foundation but wisely refrained from attending meetings of the Foundation or the Trust when decisions were made.[13]

Apart from securing its future, Wirrimbirra received other advantages as a National Trust property: taxation concessions were given on gifts to the Trust for the use of the Stead Foundation, and Trust properties are exempt from the payment of local government rates.

In 1966 the Foundation was still unsuccessfully negotiating for the lease of some land adjoining Wirrimbirra to make the area larger, and therefore more valuable as a conservation project. Correspondence had taken place between the Foundation, the Department of Lands, and the Department of Mines (since the land in question was coal-bearing). Early in 1966 an appeal was made to the Minister for Lands and Mines, the Honorable

he bus "Fez Pliz" at Careel Bay on Pittwater.

Thistle as a young girl.

David Stead

Thistle pictured at Kath Elgers place in Ryde — probably aged about 45.

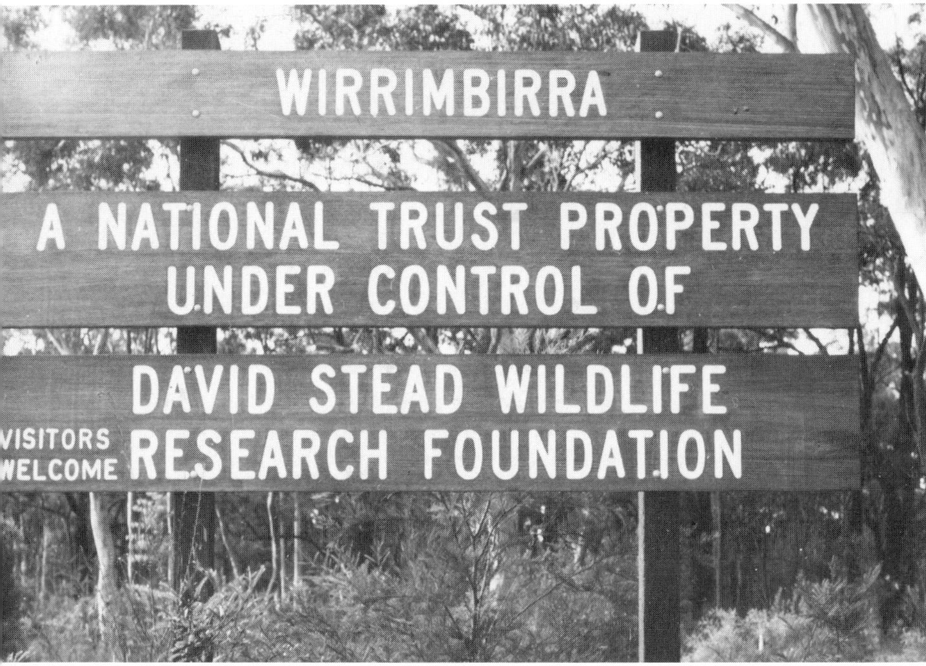

Above: *Wirrimbirra, Bargo, New South Wales.*
Below: *Native Garden named in honour of Thistle Harris at Wirrimbirra.*

Right: *Allen Strom at Wirrimbirra.*

Below: *Children studying pond life at Wirrimbirra.*

T. Lewis, MLA, and he lost no time in investigating the matter, with the result that he recommended that the Foundation be given the Trusteeship of adjoining portions 18 and 19 (78 acres approx.).[14] The application to the Department of Lands was eventually successful and the total area of nearly 200 acres came under Wirrimbirra control. In early 1972 the Foundation was informed by Mr Lewis that approval had been given for the addition of Portion 33 (35 acres) to Reserve 86335 (adjoining Wirrimbirra), under the trusteeship of the Foundation. In 1966 an application for the declaration of the area as a Wildlife Refuge was lodged with the Fauna Protection Panel (Chief Secretary's Department); the September 1967 newsletter reported that Wirrimbirra had been gazetted as Wildlife Refuge No. 163.

When the Foundation decided to erect its first building at Wirrimbirra in 1964, tenders were called. Among those received, the most attractive came from a Tahmoor firm, A. G. Miller and Son. Colin Miller, the son who had just taken over the management of the business, was visited by Thistle and Carmen, and a long association with Wirrimbirra was thus initiated. The Board members found him not only competent, reliable and generous in business, but one whose interests lay in the same direction; he became a Board member in 1972. When initiating a stand-in ranger, the advice was always, "If you have any problems, consult Colin Miller".[15] The Foundation was always mindful of the goodwill of the local residents and appreciative of the assistance it received from them. Colin Miller was well known in the local district and the high esteem in which he was held helped attract local support.

The architect for the Wirrimbirra buildings was Milo Dunphy of Loder and Dunphy, a staunch conservationist and a man of strong convictions. Colin and Milo sometimes had their differences but the end result was satisfying to the Board. "To our architect we must extend thanks for the meticulous care and thought he has put into designing buildings which will be attractive, practical and economical."[16] The first building was a small office block with a room at one end for a bed for Thistle, a bathroom, an open area at the other end for display of pot plants, and another room beyond that for tools.

Mr Mel Fackender, well known among conservationists, particularly in the Illawarra where he was an active member of the Natural History Society and at one time its President, offered his services as a Ranger in 1965. The Board gratefully accepted, and plans for the Ranger-Warden's cottage drawn up by Milo Dunphy. By March 1966 this cottage had been built by Colin Miller and Mel Fackender was in residence and on duty. Mel's first and continuing problems at Wirrimbirra involved the supply of water and even the dam built in 1965 needed good rain to fill it. Thistle wrote to a friend (Kay Geeves of Tasmania) in June 1967, "Wirrimbirra is taking a lot of time — it is beginning to be popular and while this is excellent, it is hard to keep up with". Thistle wrote to Kay Geeves again in December 1967 (having been to Tasmania with Vincent Serventy in July to support the fight to save Lake Pedder):

> "We have been in serious trouble with our architect for building 'non-homogeneous' structures at Wirrimbirra and are considering some remedial measures such as batten fences to hide the glasshouse and shadehouse. The landscaping has come in for some criticism too so, nothing daunted, I have applied for admission to the post-graduate diploma course in Landscape Design at New South Wales University. The Prof. is agreeable, if the course is held, which is still doubtful. It will be fun to see how it can be fitted in — 4 nights a week for 2 years — and how the young architects will accept great-grandmamma. But fun anyway."

By the end of 1970 a new cottage and car park were complete, the original cottage to be available for a second ranger. In June of 1970 the newsletter announced the untimely death of Carmen Coleman. She had died on March 29, 1970. Roy Ockenden, the Foundation's first secretary, had also died suddenly in early 1968, and on Saturday, October 24, 1970, friends and associates were invited to Wirrimbirra to attend the unveiling of a plaque to his memory in the Ockenden Picnic and

Parking Area, an open space with picnic bays designed by Milo Dunphy.

Another untimely death, late in 1968, was that of Barbara Dew, one of the first members of the Foundation. The Wild Life Preservation Society of Australia, on whose Council Barbara had served actively for some years, opened a memorial fund soon after her death. Barbara's mother, Lady Doreen Dew, expressed a wish that the memorial should take the form of a substantial building if adequate funds could be raised, and that the building should be erected at Wirrimbirra, in which her daughter had shown great interest, and for which she had worked very hard. The Wild Life Preservation Society, the National Trust, and the Stead Foundation all agreed that the most suitable form of this memorial should be a Field Studies Centre, thus serving Barbara's two loves, young people and natural history.[17]

School children had already been visiting Wirrimbirra for educational purposes, and volunteers had been acting as educational officers. A field studies area, with tracks designed by Allen Strom, was in use, and a secondary school assignment had been prepared by Thistle. Kath Groat was a great support for conducting primary groups around the trails, while Allen Strom and Thistle too used their expertise to good effect. Mel Fackender kept bees, and was often called out to display the contents of his hive to visiting children.

A pleasing development in the plans for a Field Studies Centre was the interest shown by the Gould League of New South Wales. In November 1970, a party of councillors from the League visited Wirrimbirra, shown around by Thistle Stead and Allen Strom. A report was presented to the Gould League members, pointing out the potential of the area and the work already done to develop the site; Thistle showed them plans for the establishment of a lecture, display and laboratory room where groups of students could be taken for lectures, examination of specimens, study, etc. Around this room it was planned to build sleeping and eating quarters for approximately twenty-five students. A meeting of the Gould League, after studying the report, passed a resolution that the League would erect a lecture, display and laboratory room for the use of visiting students at Wirrimbirra as a Gould League project.[18] The first section of the Field Studies Centre was erected with a $7,500 donation from the Gould League of New South Wales and opened in September 1971 in one large classroom catering for visiting school groups instructed by volunteers from the Foundation. This first section of the Field Studies Centre was given by the Gould League in appreciation of the long and devoted service given to the League by Mr J. E. Roberts, Secretary from 1938 to 1962. A memorial plaque was unveiled on this opening day, September 26, 1971.

> "What is an Environmental Studies Centre?" Allen Strom, President, asked those at the opening on September 26. He answered: "A place where teachers and students come together to organize for a planned and careful study of local environment — within the Sanctuary or within the surrounding districts. There are many reasons for making the study, including these:
>
> 1. To gain an appreciation of the importance of the resources of nature which we extract from our environment in order to live with enjoyment and quality;
> 2. To look at the way that nature works because these workings provide the guidelines for man's management of the resources;
> 3. To become aware of the way that man is using and abusing the resources of nature and therefore threatening the quality of life fifty, a hundred or more years ahead.
>
> Put very briefly, Environmental Studies Centres provide a means for learning the techniques involved in analysing environment and subsequently working for the benefit of man."[19]

In later years, after the Field Studies Centre had been staffed by teachers from the Department of Education, Allen Strom was critical of the programmes conducted there for many years, because they ignored the historical and social aspects of the district; he believed this was due to the fact that the teachers were science trained and so concentrated on the area with which they were familiar, "The science trained teachers never had the holistic view," he said.[20]

In 1970–71, members of the Board were Allen Strom (President), Thistle Stead (Vice-President), David Beswick, K. D. Harris, N. H. Hicks, Eunice McClure, Vincent Serventy, A. N. Thiele, I. F. Wyatt, Kathleen Groat was the Honorary Secretary, Elizabeth (Betty) Chandler was Honorary Assistant Secretary, John K. Stead was Honorary Treasurer. Allen Strom had been asked by Ivor Wyatt to take on the position of President as Bill Hicks, the first President, was away on business overseas for a substantial part of each year, and could not give the position the attention it needed. The Board, a few years previous to this, mid-year 1967, had decided to revise its rather ambitious programme devised in the early days and restrict its activities to Wirrimbirra and any other such properties it might acquire; this was in the interests of conserving their activities.[21] However, the Foundation at this time, and in the years to come, took part in other activities such as state-wide conservation campaigns. It was always represented at Annual Conferences of the Nature Conservation Council of New South Wales.

Mel Fackender was a reliable and conscientious Ranger at Wirrimbirra for a period of ten years. Charles Boyd, who first met Mel when Charles was a Field officer of the Fauna Protection Panel, remembers that first meeting:

> "My first meeting with Mel provided him with a welcome break from carrying buckets of water from a newly constructed dam, along a bush track to a newly constructed garden plot beside the cottage, in which alpine plants had been newly planted, as a result of Thistle Stead's recent return from a botanical expedition to the Snowy Mountains. I remember a Snow Daisy in flower, so it must have been summer.

"By the time my visit ended, I had no doubt about Mel's enthusiasm for nature conservation nor about his aesthetic appreciation of the natural world. Many years later I attended an Open Day at Wirrimbirra Sanctuary when Mel's ashes were sprinkled over that self-same garden bed where once he watered the wilting Snow Daisy.

"Mel generously donated his extensive collection of wildlife and environmental books to the Stead Foundation, and this is now housed in a room in the cottage as the 'Mel Fackender Library', but my memory of Mel toting those buckets of water in midsummer heat to revive those wilting plants personifies Mel's greater gift to the Wirrimbirra Sanctuary."[22]

But Mel was not always appreciated by Thistle Stead, who was frequently unsuccessful in her relationships with men. "Yes, give him a rough time she did," said Phyllis Schneider,[23] who married Mel after his retirement from Wirrimbirra, "he referred to such incidents as 'a dust-up' with Thistle. But as far as I can see, he was the life and soul of the place. Numbers of people have said to me that they never returned to Wirrimbirra after he left." Thistle was inconsistent in her attitude towards her colleagues and frequently resorted to verbal abuse.[24] Thistle and Allen Strom often had violent disagreements and when he finally resigned from the Board he said it was because he couldn't work with Thistle anymore. Colin Miller and Thistle also had their disagreements. One of her pet aversions was clearing the bush with back burning, a procedure favoured by Colin, and it was not until some buildings were nearly lost in a fierce fire in the late 70s that she relented a little. Colin said that the boundaries must be cleared, to which Thistle replied, "What's the minimum we can do?"[25]

Another problem was with trees. Milo Dunphy said you couldn't cut down trees to facilitate the building operations, but Colin couldn't get his vehicles in. One day there was an "accident" to the most troublesome tree. At one stage the nursery had too much shade, but Thistle wouldn't agree to the trees being removed. John Teulon from Swane's Nursery came to give advice and said, "If I told her what I really thought, she'd shoot me". But later on when the shade on the nursery was creating problems, she did agree to several trees being removed.[26]

One problem with Mel was recurring bouts of illness, which necessitated his being hospitalized several times. For this reason alone Thistle would have liked to dismiss Mel but Allen Strom managed to dissuade her from doing so. But it was usually Thistle who stepped into the breach and one extract from a letter to Kay Geeves in April 1969 tells just how difficult this was for her. "Mel came back on Sunday not looking very strong but you could not see my heels for dust. I had a very tough assignment for University with which both the Professor and the class members were very willing to help me but I was never able to be there to be helped. The real design section of the course is really quite beyond me. In addition to the fact that these students have had many years as practising architects or at least a number of years at drawing just these things — three at least of them are superb draughtsmen and leave me standing. However I am learning a terrific lot even if I don't make the diploma."

Throughout 1970 and 1971 Thistle kept up her hectic programme. "I'm trying to buy a hospital for an accommodation unit for our environmental studies centre, in the midst of getting ready for the Wildlife Show in Sydney and keeping Wirrimbirra going." Later she reported that "the hospital stunt didn't come off. We were afraid of damage in getting it in and were on a time limit for removal and no time to supervise. We'll get a new building, you'll see . . . The Wildlife Show took its toll, both before, during and after with a few excursions into personality clashes on the side, and a broken tooth and an extraction on my part in the middle of it all."[27]

Nothing has been said yet about the native plant nursery, which was established in 1967. In some respects it came to be the most popular part of Wirrimbirra in later years. It benefited greatly from gifts of cuttings, seedlings and seeds, not just from commercial growers, but from enthusiastic gardeners. This short extract from the annual report for the year ending December 31, 1971, gives some idea of its progress.

"Good rainfall over the summer months was a great benefit to the vegetation. Reticulated water service has been laid and mist sprays are installed in the glasshouse. Other parts of the nursery area are watered by a sprinkler system. For this work we are grateful to Mr John Teulon. For most of the year a nurseryman, Mr Brockenshire, has been employed for five days a week, and sometimes another man part-time. The plants in the nursery are in fine condition and sales have expanded to the extent that the nursery is self-supporting."[28]

But the picture was not always as pleasing as this, as Thistle reported in her letters to her Tasmanian friend, Kay Geeves:

5.6.67 It is raining N. and S. of us but not a drop here.

20.8.68 Just back from W. — still drought and bad frost. Lost hundreds of plants.

18.10.68 Things are as bad as ever, but rain will come one day and we will start again.

30.10.69 W. is recovering rapidly and looks fine.

23.11.69 We had flood rains at W. but I haven't been down to see how much damage has been done. Also some difficulties with the Ranger who is becoming a bit big-headed being there in complete charge so much.

6.7.70 Terribly dry at Wirrimbirra again — our dams are very low indeed.

21.2.71 The Ranger's report indicated losses in the glasshouse due to excessive heat (up to 106°) but I think proper attention could have saved them. (Thistle herself had been away in Tasmania.)

1.3.71 Not enough rain at W. but overcast weather and humidity plus much bad management has caused devastation — most of the glasshouse material lost completely and a lot of the other potted stuff.

14.8.80 Winter was very cold, and following the long drought — it is still very dry — the spring will be poor for us.

31.12.86 We had a disastrous hailstorm at W. a few weeks ago. Polyhouse and sarlon house on the ground, glasshouse badly smashed, stock plants all to be repotted, and the plantations knocked about.

The Board members, Ranger, and Thistle herself were all essential parts of Wirrimbirra's management, but in the early days, in fact for the first two decades, an enormous amount of support came from volunteers, many of them Thistle's personal friends. It is impossible in this account to name all those willing people who helped Wirrimbirra grow and survive, but those who are mentioned below are only named in order to illustrate the wide range of activities in which the volunteers were involved. In *Wildlife Research News* March 1967, a vote of thanks was given to Mr Bill Keech, who organized the erection of the bush house, Mr Don Ashford who, over several weekends, helped mark out the boundary of the property and who sewed the sarlon cover of the bush house, Mrs Lois Sourry for the donation of plants and work on the Alpine Garden, Mr John Stead who helped to place stepping stones on the top wall of the dam. Throughout the seventies John Stead also looked after Wirrimbirra's financial matters, only relinquishing his position as Honorary Treasurer when he started work in Queensland in 1980. In expressions of appreciation to the volunteers, Mrs Ivy Stead was often remembered for her supervision of afternoon tea at meetings, as well as planting, and lettering and painting labels. In March 1968 gifts were acknowledged from Brother Pat Stanley (plants), Don Perrin (a supply of printed seed envelopes and copies of a useful set of notes on propagation), Mel Fackender (a valuable collection of workshop and garden tools), Nancy Jeffery (paintings of the nine eastern species of *Banksia*, to be placed in the display pavilion). In December 1968, useful gifts were jerricans of water, as Wirrimbirra was in the throes of a drought. An appeal was made for donations to help in the purchase of water. At this particular time, a second dam built in

1967 had never had water. In 1969 a Water Fund was opened to attract finance for installing a reliable water supply from the town mains. This water was eventually connected at Easter 1971 and made an enormous difference for keeping up a water supply to the plants.

Back in 1964, Mrs Lois Sourry had offered to finance the publication of 1,000 copies of a booklet on the cultivation of Australian plants. The offer was accepted and the text prepared. In 1970 Miss Jean Stevenson and Miss Nell Reeve donated fly screens for the cottage; John Clarke donated a carpet; members of the Australian Wildlife Club (a youth group) gave a weekend of work; Betty McMahon came to the rescue with the March news sheet. Mrs Elizabeth (Betty) Chandler, as Honorary Assistant Secretary, was also Minutes Secretary and convenor for the Acacia Study Group. "Betty is the most consistent of all members in helping with the task of keeping Wirrimbirra flourishing, as anyone who visits Wirrimbirra regularly well knows. She is excellent with potting, weeding implements and secateurs, often helps out with sales in the nursery and acts as general roustabout."[29] Elizabeth was also responsible for a publication on 200 Acacias.

In expressing appreciation for volunteers, those who undertook the arduous task of Editor of *Wildlife Research News* deserve a special mention. In 1970, Rosemary Brown, who edited the news sheet for all issues except No. 1 in 1964, up to 1970, found it necessary to resign, and the position filled by Dora Reiach and Rita Barnes. At a later date, Mary Crocker fulfilled the role for some time, until relieved by Susan Heins. In 1971 the *News* took on a new format due to the efforts of Betty and Gilbert Stead, the Foundation's unofficial archivists.

Different members of the Stead family were great supporters of Thistle's project, right from its inception. John Stead and his parents, Ivy and Kel, who all lived at Boongarre, often went to Wirrimbirra with Thistle in the sixties and seventies, and she would offer to supply the lunch, her speciality, a curry. "Thistle could make a pretty reasonable curry, but it was always burnt," said John. "If we were going bushwalking or to Wirrimbirra for the day she would inevitably prepare a curry for lunch and it would be in a billy which could be hung over a fire or put on a camp stove but it was never stirred and consequently the bottom of the stew became a layer of black flakes. As ants often made their way into the container prior to its being heated and eaten, Thistle's curries took on a whispered cult status."[30]

Research had been an important objective in the early days, but involvement in this area became less as time went on and finance and volunteers were in short supply. In 1965 an agreement was signed between the Foundation and Mr W. R. Keech, well known in the Oatley Fauna and Flora Protection Society. Under this agreement, Mr Keech was leased approximately two and a half acres of Block 17, at the eastern end of which he planned to grow a wide variety of *Grevillea* species. In June 1969 the newsletter reported that Bill Keech, assisted by a University student, Peter Clifton, planted nearly 85 grevilleas, comprising 25 species and 6 different hybrids. Mel Fackender helped them make a long siphon line from the dam to provide water to the area. Another early piece of research, a bird dialect study, was undertaken by Mr E. W. Chandler. Research in the growth and propagation of Australian native plants was also an integral part of the work of the nursery.

In 1972 Don Perrin, a former teacher, joined the teaching team at Wirrimbirra on a voluntary basis, and before the year was out, two new buildings had been erected in the Dew Building Complex for the Field Studies Centre — a toilet/ablution block to service up to 40 resident students, and the first of the 8-bed dormitories. Others were to follow, with assistance from various groups. For example, the Persoonia cabin was built with financial support from CSR (Colonial Sugar Refining Company) and opened on May 25, 1974; the Banksia cabin was built with funds from the Wild Life Preservation Society and opened on May 24, 1975. Colin Miller continued to be the builder for all the new work. By the end of 1972 the Allen Strom Demonstration Natural Area and pool close to the classroom was well established; this area was not only for use by visiting students, but planned to be a model to assist teachers in the design and use of their own areas. Allen Strom wrote an article for *Wildlife Research News* in September 1972, in which he said:

> "It needs more than a brief four hours' visit, once in a child's school life, to make much of an impression on his attitudes and future conduct. It will soon be possible to extend the time of visits and give a more prolonged exposure to the experience of being in the bush. Even so we should like to see return visits by children who have already been there . . . In real fact, it is the teachers that matter. The handful of people in the Wirrimbirra Teaching Team can never hope to affect many of the children of today, but the teachers are a more reasonable proposition . . . to think of Wirrimbirra providing the guidelines for teachers to work on their own with their classes around the school grounds, around the school district and in places wherever the teacher chooses is much nearer to being a feasible approach to spreading environmental teaching."

Reference should be made to Thistle's attitude to the "capitalist world" because in at least one instance it did lead to a loss of support for Wirrimbirra. John Stead, who did not follow Thistle's socialist politics, had many disagreements with her about the rights and wrongs of political systems and the roles of business corporations. CSR, as mentioned above, gave financial assistance for the construction of one of the cabins, but later declined to continue with their financial support. Thistle, with others, was pursuing the anti-logging/woodchip debate at this time and CSR was involved in the issue also. Whether Thistle's stand on the matter was a factor in the CSR decision about Wirrimbirra is not clear, says John, but he did point out to Thistle at the time that you shouldn't bite the hand that feeds you! Thistle's attitude, however, was that one should not allow corporations to throw a few dollars to charity to buy off their evils.[31]

The Stead Foundation announced in its December 1972 issue of *Wildlife Research News* its full support for a new organization established at a seminar and public meeting at Macquarie University on September 8-9, 1972. This was the Association for Environmental Education (New South Wales), and Allen Strom, Adviser in Conservation (Department of Education New South Wales), and President of the Stead Wildlife Foundation, was elected Vice-president, along with Professor Ian Douglas, Department of Geography, University of New England; Thistle was elected to the Committee, while Dr G. R. (Rex) Meyer, the Director of the Centre for the Advancement of Teaching at Macquarie University, was elected President.

The Stead Foundation Board had nominated its President, Allen Strom, for the Australian Natural History Medallion for 1972, and was gratified when he was the worthy recipient. The medallion is awarded annually by the Field Naturalists' Club of Victoria.

On April 16, 1973 came news that the Field Studies Centre school, with a teacher from the Department of Education, would be in operation within a few months. The school, designated a School for Specific Purpose, was staffed by Mr Keith Armstrong, previously on the staff of Sylvania High School. Public school teachers were advised to contact Mr Armstrong for bookings; classes would be held from Monday to Thursday inclusive, except during school holidays. Private schools or other groups wishing to use the facilities still had to contact the Convenor of the Foundation Teaching Team, Kath Groat. Mr C. L. Macdonald, the Inspector of Schools in charge of Environmental Education, predicted that Wirrimbirra would be the pattern for the establishment of similar schools throughout the State. The school was officially opened by the Honorable A. E. Willis, Minister for Education, on Monday, August 20, 1973, the second such school to be opened in New South Wales, the first being at Muogamarra Sanctuary, Cowan.

Co-operation was the theme common to the speeches given at the opening of Wirrimbirra Field Studies Centre school: the property had been transferred to the National Trust, but was managed by the Foundation; the Field Studies Centre was built by the Gould League but furnished and equipped by the Department of Education; teachers and students, group leaders and members of groups eagerly participated when the Barbara Dew Field Studies Centre began operating — with a voluntary but well qualified staff. So Wirrimbirra was a going concern by the time the Department of Education was ready to establish a school. The ceremony also marked the recognition of Wirrimbirra as a part of the Bargo District community. Members of local social services were present as well as distinguished representatives of local government and of the area directorate of education.

From the first day that Keith Armstrong took up his position in April 1973 at the Field Studies Centre, Thistle was there to initiate him thoroughly into the history and mission of the Stead Foundation. Keith had only met Thistle once previously, at the interview in Bridge Street, Head Office of the Department of Education, when Inspector Colin Macdonald and Thistle were checking the applicants for the position at Wirrimbirra. Keith wondered who "this little old lady" was, and felt at the end of the interview that he was not her choice. She told him later that he was her second choice.[32]

As someone with very little experience in native plants and conservation education, Keith went along with Thistle in whatever she was showing him or talking about. "She gave me a lot of her time and expertise which I really appreciated. She even loaned me money to buy a block of land in order to get my family settled and prepared to stay."[33] But when all of Keith's equipment had arrived, a divergence in their thinking over certain matters began to develop. Thistle thought teaching resources were best stored out of sight, while Keith believed that children and staff needed to walk into a stimulating place with a variety of things to interest them. "It's all too messy Keith," she would say. She would watch while Keith conducted a teaching session, and there were quite intense discussions afterwards about their different teaching methods. Keith's groups were large (40 students) with a variety of experience and interest and he was committed to "turning them on" to the bush by guided exploration of the environment without Thistle's feverish urge to teach them identification and classification of native plants. "We had our differences until I invited her to take a group one day and show me how — she found it very difficult and decided to leave the work with the kids to me!"[34]

Thistle was always trying to get money from visiting groups, Keith recalls, and she insisted he take the school groups to the nursery to buy plants, or posters, etc. to take home. She always had a good supply of the tree *Grevillea robusta* available because "they were easy to germinate and the kids can't kill them". Keith still has visions of all those giant trees growing in the window boxes of high rise units in Liverpool.

Keith Armstrong was not all that impressed with the Wirrimbirra site. "The 100 hectares was pretty terrible sandstone scrub with a few hanging swamps and a moist gully through it." Vincent Serventy too expressed the opinion that "it was a terrible piece of land, there was much better land closer to Sydney". Keith was using a series of sites along already established tracks, but these became overused fairly rapidly, and he wisely moved things around regularly. This didn't impress Thistle as she failed to accept that there had to be some damage with the numbers using the area.

At the end of 1978 Keith Armstrong left Wirrimbirra to return to mainstream schooling, becoming Head Teacher in Science at Canobolas High School in Orange in 1979. He left with mixed feelings, having benefited enormously from his association with Thistle, Allen Strom, and Mel Fackender. "Professionally I developed through my dealings with students and staff from kindergarten to university. I was proud that I had played a part in starting the Centre and that the number of Centres in New South Wales had grown from 2 to 13 during the time."[35] In *Wildlife Research News* for June 1979 Allen Strom wrote this tribute to Keith Armstrong. "In all Armstrong showed considerable adaptability and spontaneity in an entirely new field of education. He was excellent to work with; he was remarkably understanding of the management problems of Wirrimbirra. He presented Wirrimbirra as a concept in nature conservation both within the local community where he was highly regarded, and among the teaching profession. Altogether he made an outstanding contribution to environmental education."

In October 1973 *Wildlife Research News* reported on "A Bridge over a Troubled Railway". Some years previously the Stead Foundation sought permission to cross the Great Southern Railway from the administrative area to the study areas. This was refused on safety grounds so it was necessary to go south towards Bargo to cross the railway on an overhead bridge and then double back — a distance of about four kilometres. Now that the school was in operation, the travelling was found to be even more inconvenient, so the Foundation asked the Public Transport Commission to consider the erection of a footbridge. The Commission agreed to this, but the Foundation was to bear the expense of construction at twenty thousand dollars. The Board said it wanted a bridge, but there were other more pressing needs to be met. So the troubled railway remained unbridged.

During 1973 the Foundation lost three of its valuable workers, George Duguid, one of the nurserymen and a reliable assistant, died early in the year. Loretto Williams, who was Minutes Secretary and Editor of *Your Australian Garden* booklets died after a long and distressing illness. Gwen Hanley, who helped in innumerable ways and was most enthusiastic over all Wirrimbirra projects, died suddenly during the year. Plans for memorials to Loretto and Gwen were in hand.

Thistle wrote to Kay Geeves early in 1973, "We have a ranger complete with charming wife and 14 year old daughter taking up residence at W. with a view to permanancy should we each like each other, which I think we will". This was F. K. S. Woods, O.B.E., who had been Principal of the Fairbridge Farm Schools near Molong, New South Wales, for 27 years, and the position was to act as assistant and support to Mel Fackender. However, he only stayed a few months, and the Foundation was unsuccessful in getting a second ranger until early in 1974, when Robert Jones, a graduate from Macquarie University, was appointed as Assistant Ranger. Unfortunately he too only stayed a few months. In March 1975 Bruce Knight was appointed as a ranger, but owing to the Foundation's precarious financial situation, his appointment was for one year only.

Mel Fackender was to leave at the end of 1975 with a move to Adelaide to marry Phyllis Symon. His actual departure from Wirrimbirra was in January 1976 because he wished to complete his ten years as Ranger at the Sanctuary. The marriage of Mel and Phyllis was a rewarding one for both of them. Phyllis cherished Mel and under her care his health improved greatly. However, after a few happy years Mel died in Adelaide on October 1, 1982 at the age of 81. Many were the tributes expressed at his funeral, and at the interring of his ashes at Wirrimbirra on October 16. Dr Brian Symon, Mel's step-son, spoke at the funeral saying in part, "The major impression created by Mel was that of a lover of life. Not his own life. In fact, in the time I knew him I cannot recall a single self-interested statement. He took a totally satisfied and uncomplaining attitude to his own environment. Mel's major interest was the beauty of life around us. He was acutely conscious of the treasures the world has to offer and also the speed with which the heritage of our children and grandchildren is being destroyed. To say this alone however, would be to give a totally superficial view of Mel's commitment to action. Mel had a talented intellect which he himself educated to an uncommonly high level. His intellect was matched with a powerful memory. He was capable of achieving an iron grip on literary pieces which he thought of significance and would quote most aptly to enliven any conversation."

Three years after Mel Fackender's death in 1982, his widow Phyllis published a booklet as a tribute to his memory, including in its pages eulogies and reflections on his life written by friends and associates. Her son's tribute above was included in the booklet. On October 19, 1982 about 50 people gathered for a Memorial Service at Wollongong Botanic Gardens. One of the speakers was Norm Robinson, former President and in 1982,

Vice-president of the Illawarra Natural History Society, of which Mel had been a life member. Norm said, "Wirrimbirra Study Centre is a reminder of his dedication to conservation and Natural history. Countless native shrubs, wildflowers and many native trees were grown by him. In a world full of greed, Mel showed his true self in working for years for no financial rewards whatsoever at this Centre. It is a great pity that the world contains so few people of his calibre. I am pleased to have been associated with him during his lifetime." Keith Armstrong, the first teacher of the Wirrimbirra Field Studies Centre, in reply to a query as to whether he had come all the way from Orange for Mel's ceremony, said, "Yes, I did, because he was the best teacher I've ever had and I'll be grateful to him to the end of my days".

Allen Strom had known Mel Fackender for a period of over forty years. He wrote a tribute especially for the booklet produced by Phyllis. Allen concluded his tribute by saying, "Mel Fackender had his own standards of response to the innumerable demands of living at Wirrimbirra and being responsible for long hours of loneliness of management — a management for which there were few guidelines and few people to provide advice. That he could succeed in bringing Wirrimbirra from a pot pourri of nebulous thoughts to an establishment of some consequence, must be sufficient epitaph. Like all pioneers however, his value was most evident in profile some years ahead of his time. Mel worked at Wirrimbirra because he believed in the concept of preservation of naturalness and the encouragement of human endeavour to understand the value of that naturalness. The odds against him, physical and human, were considerable. I often wondered how he kept going — perhaps it was the appreciation of those who came from time to time to his temple."

Chapter 7 Notes

1. *Australian Wild Life*, Vol. 3, No. 3, March 1958, p. 7.
2. Unpublished notes by T. Stead.
3. Land Titles Office.
4. Unpublished notes by T. Stead.

5. Land Titles Office.
6. Leaflet produced by David G. Stead Memorial Wildlife Research Foundation.
7. ibid.
8. I. Wyatt, pers. comm. 16.1.96.
9. *Wildlife Research News*, No. 2, June 1964, p. 4.
10. ibid. No. 26, June 1970, p. 3.
11. ibid. No. 3, September 1964, p. 2.
12. ibid. No. 9, March 1996, p. 4.
13. I. Wyatt, pers. comm. 28.4.97.
14. *Wildlife Research News*, No. 9, March 1966, p. 5.
15. ibid. Vol. 2, No. 12, December 1973, p. 11.
16. ibid. No. 1, March 1964, p. 6.
17. ibid. No. 25, March 1970, p. 2.
18. ibid. No. 28, December 1970, p. 13.
19. ibid. Vol. 2, No. 4, December 1971, p. 5.
20. A. Strom, pers. comm. 2.1.97.
21. Letter by T. Stead to K. Geeves, 5.6.67.
22. C. Boyd in *Wildlife Research News*, October 1993.
23. Phyllis Fackender remarried and is now Phyllis Schneider.
24. A. Strom, pers. comm. 2.1.97.
25. C. Miller, pers. comm. 17.4.97.
26. ibid.
27. Letter by T. Stead to K. Geeves, 1.9.71.
28. *Wildlife Research News*, Vol. 2, No. 5, March 1972.
29. ibid. Vol. 3, No. 2, June 1974, p. 11.
30. J. Stead, pers. com. 14.7.97.
31. ibid.
32. K. Armstrong, pers. comm. by letter, 14.5.97.
33. ibid.
34. ibid.
35. ibid.

CHAPTER 8

A Hectic Life — 1964–1969

"The poor old Alpine Plants book is getting short shrift again."

T. Y. Stead, 1967

IN the early to mid sixties the Wild Life Preservation Society of Australia was passing through the doldrums. *Australian Wild Life*, the Journal of the Society, carried an editorial in its June 1965 issue as part of its report covering the period from March 1964 to February 1965. It is here reproduced in part:

> "The Society cannot look with much pride on the year's achievements. While it is true that much work has been done and many things accomplished, they have almost entirely failed to sell the Society's principles to the community. The Society has a duty to the public, particularly at the level of education, which simply is not being fulfilled.
>
> "Almost twelve months ago attempts were made to set up an Education Committee to get our message across to schools and young people. Not a single member of the Society was prepared to organize the Committee which has consequently not yet been convened . . .
>
> "Our approach to juvenile education so far has been much like harvesting ten thousand acres of grain with a pen knife . . .
>
> "Adult education is not seriously undertaken by the Society . . . Our public relations are not strong enough . . . In summary this Society cannot rest until it has helped to promote an attitude in the public mind which accepts wild

life preservation as just as much part of daily life as driving on the left of the road.

"To do this we must refurbish our rather faded public image and give Australia the idea that we are doing something . . .

"An artificial increase of members and funds followed the amalgamation of the Naturalists' Society during the year."[1]

This introduction to the Report for 1964 was signed by J. M. Bennett, President, but it must have had the consent and approval of the Honorary Editor, Thistle Stead. From 1966, however, matters started to improve for the Society because Vincent Serventy became President, and he brought in new ideas and a fresh approach. Vincent was a resident of Western Australia until 1965 when he came to live in New South Wales. Always a keen naturalist, he had joined the Wild Life Preservation Society in Western Australia in the forties, and also met Thistle in the forties when on a visit to New South Wales. She took Vincent around for a week — they tramped in Ku-ring-gai Chase National Park, the Botanic Gardens, and Centennial Park, and had a great time, Vincent recalls. They got on well and he was impressed with Thistle's enthusiasm. However, Thistle had changed since the death of David Stead, and in her relationships with other people she was likely to be quick-tempered and at times even disagreeable. Two earnest people on the Council of the Society, Jim and Mabel Brown, invited Vincent Serventy to dinner one evening and shared with him their misgivings about the state of the Society, the fact that Thistle's often abusive manner was driving Council members away. Jim Brown asked Vincent to stand for President, and he agreed. Mr J. M. Bennett resigned in July 1966 and Vincent Serventy became President, a position he has held until the present, but for one year. Thistle became Vice-President, one of several, and held that position for many years.[2]

Vincent and Thistle worked agreeably together for the most part, but every now and then there were outbursts, and the Council continued to shrink because its members didn't want to face the squabbles. On one matter at least Vincent and Thistle were in opposite camps; he had been associated with the mutton bird industry, and believed that culling of kangaroos, done in a humane way, was sometimes necessary to deal with population increase. But Thistle saw things as black and white, there was no compromise for her; there could be no killing at all, and the commercial aspect of wild life was not to be considered on any terms. Vincent believe that to work with people effectively there had to be compromise now and then.[3]

Meanwhile Thistle had developed another interest, in addition to the growing responsibility and involvement with Wirrimbirra. She had been working on another book, about alpine flora, and decided to go to Tasmania to visit the alpine areas there. On this trip she was accompanied by Eve Gourlay who was lecturing in Biology at Sydney Teachers' College, being seconded from 1960 to 1967. It was December 1964, and the two friends set out for Melbourne in Thistle's car, a tour of ten weeks having been planned for them by Jack Thwaites, Superintendent of the Scenery Preservation Board of Tasmania. Jack was also a foundation member of the Hobart Walking Club and the Youth Hostels Association of Tasmania, and was a delightful man, his Tasmanian friends recall. He had organized the accommodation for Thistle and Eve — sometimes a Forestry hut, or a cabin high in the mountains, or in a private home of Jack's friends, as in Launceston.[4]

They travelled across Bass Strait in the ferry and travelled first down the west coast to Rosebery, where it rained incessantly, then to Queenstown and down to Macquarie Harbour, and across to Hobart. Eve recalls Thistle with more than one camera around her neck at one time, meticulous about the light as she fussed about her photographs, taking hours for each one, it seemed to Eve. Late in December Jack Thwaites took them both to Hartz National Park while they were staying in Hobart, and one evening Thistle gave a slide presentation of Australian native plants. After her lecture, a young woman came up to Thistle to tell her how much she'd enjoyed the presentation, introducing herself as Kay Geeves from Port Huon, not far from Hartz Mountain. Oh, said Thistle, I'll be down your way tomorrow, I left my lens cap at Hartz, I'll call in to see you. Thus began a friendship and a correspondence by letter that lasted for 25 years until Thistle's death.[5]

At that time, 1964, there was no branch of the SGAP in Tasmania; the south-east region of SGAP included Victoria and Tasmania, and Kay Geeves belonged to that group. She was also on the South-West Committee at that time, a non-government body of conservation-conscious representatives of various walking clubs and associations, the main concern at this time being rumoured threats about the Gordon River and Lake Pedder. Thistle told Kay she believed a branch of the SGAP should be set up in Tasmania, and encouraged Kay to find some like-minded people to share her enthusiasm. This initial encouragement was not to falter. Writing to the author in February 1997, Kay said, "In reading them [Thistle's letters] again I am struck by her tremendous vitality, enthusiasm and output in the sixties and seventies. Her optimism, persistence and dedication were inspiring. These were momentous times in Tasmania's developing conservation movement. My calls for help to Thistle were always answered a hundredfold. She had the experience we lacked and was enormously generous with her time and knowledge. She advised us, gave us national and international contacts and poured out letters on Lake Pedder and the south-west and numerous other issues."[6]

From Hobart, Eve and Thistle travelled north to Launceston, Eve doing most of the driving throughout the whole trip, as Thistle was at the ready with her cameras. Then it was to Cradle Mountain, being caught in snow in early January, and back to Hobart, followed by an unforgettable trip to Lake Pedder. "Access is by foot or by light plane," Thistle wrote in a journal article.[7] "We chose the plane, a three-passenger Cessna. 'Aren't you scairt?' asked a young German friend. To my bold denial she replied coyly — 'Just a liddl' bit scairt.' And truly, as the wee plane shivered over deep and remote gorges, I understood what she meant.

"But it was a never-to-be-forgotten flight with a pilot who, like all who have visited Pedder, loved the place, pointing out the various peaks and ranges over which we passed. Landing on the quartzite beach, we found ourselves on the margin of the shallow glacial lake, its brownish buttongrass-coloured waters lapping gently along a completely deserted shore with Mount Anne rising serenely behind Mount Solitary only partly visible through a low cloud and the Frankland Range beckoning an invitation to explore. As we saw the lake and its encircling mountains in their various moods we came to understand the walker's love for this place. Even the buttongrass plains, with their treacherously concealed bogs and deep waterholes, could not keep us from an attempted climb to the top of the range.

"At Port Davey I found the last of the season's buds of *Hewardia tasmanica* in Mrs Clyde Clayton's garden . . . On Melaleuca Inlet we camped in the comfortable King Memorial Hut and explored the many bays of Port Davey and Bathurst Harbour . . . The Scenery Preservation Board maintains a fire-watching service on the reserve during the summer months to protect this remote and wild country."[8] Eve remembers being caught in snow on Mt Rugby. This is a rugged little peak (771 metres) adjacent to the Bathurst Narrows at Port Davey.

Back in Hobart there was another visit to Kay Geeves. From Hobart they travelled up the east coast, visited Freycinet Peninsula National Park, and back up north to catch the ferry to the mainland.

Thistle was very keen for Kay Geeves to meet a fellow Tasmanian, Alan Gray, with whom she had been corresponding since the early sixties. She had seen Alan's articles on Tasmanian flora in the *Australian Plants Journal*, a magazine published on behalf of the Society for Growing Australian Plants and its member societies. The articles were concerned with the description and propagation of the then obscurely recognized Tasmanian endemic and alpine plants. Thistle wrote to Alan asking for further information, photographs, and living plant material, having in mind her book on alpine flora which she was researching.

Alan was to eventually meet Thistle at her lecture in Hobart in December 1964 but he did not know Kay Geeves at that time.

In February 1967 when disastrous fires swept through the environs of Hobart, Thistle was writing concerned letters to check on her new friends. Alan Gray's house did burn down in those fires. Through Thistle, Kay Geeves eventually met Alan Gray, and in November 1968 Kay organized the first meeting, which fifteen people attended, and at which the decision was made to form a local Tasmanian group of SGAP. Two years later this became a Region in its own right. The Tasmanian members give Thistle the credit for the encouragement to start this group, "she was a great networker, a catalyst".

In June 1965, *Australian Wild Life*, the journal of the Wild Life Preservation Society, published an article by Thistle about Tasmania, entitled "Island of Enchantment". "No part of Australia has more to offer by way of scenic beauty than Tasmania; nor, perhaps, more of scientific interest. A considerable mineral wealth, ancient volcanic peaks, evidence of glaciation in sweeping valleys, lakes and tarns, much of it still incompletely described, offers a challenge to the geologist. Its endemic flora, not large in number of species, is sufficiently different from the rest of the flora of Australia and of such beauty, as to provide botanists with continual delight."[9] The island south of Bass Strait had captured Thistle's interest and concern, as she discussed the alienation of much of its land, and the threat to much more. "And what are the surveyors in their helicopters over Lake Pedder planning as they take bearings — must this lovely lake suffer the fate of becoming a storage dam instead of remaining a naturalists' paradise? . . . Nor are the reserves safe from exploitation . . . The motorist has a right to enjoyment of natural beauty and the conservationist would not deny this . . . But planning and moderation can still result in happy compromise. The motorist must appreciate that some parts will ever remain outside his compass; the conservationist must realize that adequate playing areas and picnic grounds are an essential part of modern civilized living and a valuable buffer to wilderness areas if properly managed."[10]

In 1966 Thistle was the author of a booklet *Australian Plant Life*, part of the Around Australia Program. Writing about "The Heath Family" on page 58, she said, "But of all the heaths, the creeping fuchsia, *Prionotes cerinthoides*, from the temperate rainforests of Tasmania, is perhaps the loveliest. Its thin stems twine around the tall trees and, in late summer, produce great masses of large, inflated, waxy-looking, delicate deep-pink or red bells."

During the mid-sixties, Thistle had been Editor of a series of small books, *Australian Nature Series* for 7–9 year olds, and for 10–12 year olds. She encouraged all her friends to contribute a booklet, to write on their area of interest and expertise. The series, published by Longmans, Green and Co. Ltd., Victoria, had its first booklet published in 1965, each one thirty-two pages, and illustrated with coloured drawings. Eva S. Levis wrote "Leaves"; Pat Bourke wrote "Birds in Your Garden" and "Bird Migration"; Lois Sourry wrote "Saving our Wildflowers" and "Saving our Wildlife"; Harold Cogger was responsible for "Animals and their Young" and another one on "Reptiles"; Thistle wrote "Plants Without Flowers". The Junior series, 7–9 years, had eight titles, and the Senior series, 10–12 years, had ten titles, the aim of the series being to encourage children to observe and to interpret nature.

Lois Sourry, also a keen photographer of native flora, was planning to write a book on the Hawkesbury sandstone flora, and during this time, the late sixties to early seventies, was preparing the many slides to be used for the illustrations. But Thistle threw cold water on the idea and would give Lois no encouragement, a great disappointment to Lois. Thistle could not tolerate any competition, and it was probably from around the early to mid seventies, that the close friendship between Lois and Thistle began to break down. In a sense, they drifted apart, because although with much in common, they had different commitments; Thistle was busy with Wirrimbirra and wanted Lois to spend more time there, but Lois wanted to concentrate on local issues of the Central Coast. Lois Sourry died in 1987.[11]

"That's a Thistle tree," Andrew Sourry said when the author visited his home in November 1996. "It's the Magenta Lillypilly, *Syzygium paniculatum*, which she propagated from seed at Watson's Bay. We built the house in 1958 and she gave us six plants in 1959. It was later to become an endangered species and has now been reclassified as vulnerable." Lois and Andrew kept two of the trees and gave the other four to local schools under the auspices of the Gosford Fauna and Flora Protection Society, of which Lois was President. Lois went with Thistle to Kosciuszko many times in the years 1955–1965, as already mentioned in Chapter 6, but they also travelled together on many other trips; several times they went to see Bill Cane at Mafra in Victoria — he was one of the early growers of native plants; Lois and Andrew and their daughter, Marika, went with Thistle to the Grampians (in two cars), to Dorrigo National Park, again in two cars — Thistle would sleep in hers — and another trip out to Broken Hill and the Menindee Lakes. Thistle would visit the Broken Hill Zinc Corporation to give them advice and encouragement to grow native plants in their gardens, following on the work of Bert Morris many years before. Yet another trip for the Sourry family with Thistle was to Quoin Island, out from Gladstone, probably very early seventies. She was very good company, Andrew said.[12]

Marika remembers being a child of about eight years old, sitting in the car between Lois and Thistle, about to take off on a trip. Marika says to Thistle, "I won't go with you until you put both hands on the wheel and stop smoking". Thistle could be demanding — she once sent eight-year old Marika across an icy stream to collect a specimen. In 1972 Marika, then a young woman, represented her mother, Lois Sourry, at an author's dinner at the University of Sydney. Thistle was there, and Marika went to talk to her, surprised to see her in a fur coat, not fox or rabbit, she recalls. Thistle's family says she did not own a fur coat, but if she had borrowed it, it was still an unexpected thing for her to do, a complete antithesis to what she claimed; she was a complex person, Marika said.[13]

In 1967 a battle was being fought by the North Entrance Peninsula Preservation Society on the Central Coast of New South Wales (NEPPS) to preserve as much as possible of the littoral rainforest between North Entrance and Toukley from rutile mining. The area was also forested with Sydney Red Gum, *Angophora costata.* Lois Sourry invited Vincent Serventy and Thistle Stead to visit the area and talk to the Society. Three hundred people attended a meeting beneath a huge Sydney Red Gum on June 28, 1967. Thistle spoke on conservation as Vice-President of the Wild Life Preservation Society of Australia. She appealed for everyone's support in "a combined effort to determine future land usage in this State, that due regard may be had for the maintenance of recreation areas and wild life sanctuaries". She said the rutile companies were compelled by law to take steps to restore the vegetation. "Don't be fooled by their statements that they can put things back as good as or even better than they found them," she added.[14]

The campaign was successful because the area in question was not subjected to mining, which was restricted to the ocean side. This area of significant littoral rainforest and red gum forest has since become part of Wyrrabalong National Park, which was proclaimed in 1991.

Meanwhile matters had come to a head in Tasmania where, in April 1965, proposals for the flooding of Lake Pedder had been outlined by the Premier. Road works were already well advanced and obviously following some definite plan. In June 1966 an approach was made to Federal Parliament for $47 million for the Gordon River scheme, but the granting of this sum was not announced until the day that the Middle Gordon scheme received parliamentary approval. In March 1967, the Save Lake Pedder National Park Committee was formed, but by May the Middle Gordon scheme had been tabled in Parliament and the details made public. A petition disapproving of the scheme was circulated and 8,500 signatures obtained, but this petition was not mentioned in Parliament.

Thistle, in Sydney, was mobilized for the Lake Pedder emergency. She rang various people in Sydney, sent a telegram to the Premier of Tasmania, and wrote a long letter to the Editor of the *Australian* entitled "Save Lake Pedder", under the banner of the Wild Life Preservation Society of Australia. She pointed out in this letter that the South-West Committee (formed in November 1962) was campaigning for a large national park in the area, to include Pedder, and which would have its own authority for development. This proposal had received support not only from Tasmanians, but from many individuals and organizations in most, if not all, mainland States. Last week, she claims, it was reported in the Hobart *Mercury* of 31.5.67 that an inter-departmental committee set up to report on the south-west had advised against the South-West Committee's recommendations and had recommended instead the creation of a new scenic reserve which would exclude Lake Pedder. The report, the *Mercury* claimed, maintained that the loss of Lake Pedder would not be of serious consequence. She concludes this letter to the press by saying, "We on the mainland who appreciate the beauty of this little island and who have some guilty feelings for the despoliation of so much of our own natural beauty might well feel disposed to make public our wish that Lake Pedder should not be sacrificed in the interests of economy. It would be a gesture of recognition of the proper pride Tasmania has always shown in the natural beauty of her country and a tribute to the South-West Committee which has worked so hard in the interests of the nature lovers and conservationists."[15]

It was about mid-June 1967 that Thistle and Vincent Serventy decided to go to Tasmania to lend their support to the "Save Lake Pedder" campaign, certain that news of mainlanders going down to Tasmania for this purpose would make a more dramatic strike for publicity than anything else so far. Thistle wrote to Kay Geeves to assure her that Vincent "could put on a first rate show with excellent slides, which would be first rate publicity as well as money making. We would both be happy to talk in Hobart and Launceston and Devonport if it could be arranged in the time. I think I could give good material on the unique botanical features of Pedder as an example of environmental habitat studies since crammed into this relatively small area you have probably every habitat that occurs in Tasmania. I can't honestly say that there are species in the area that occur nowhere else, at least to my knowledge, but I can stress the relative abundance of some rare species such as *Hewardia*."[16] In the same letter Thistle extolled the virtues of Vincent Serventy, no doubt to reassure Kay and her supporters that organising public meetings for both Vincent and herself would not be a fruitless exercise. "I should have added that although Vincent has not visited Pedder he knows other parts of Tas. well and has a good general knowledge of National Parks and the National Park concept, is a most convincing speaker, unemotional but enthusiastic, and with a personality that seems to attract everybody. As the author of *A continent in danger* he has prestige in this field."[17]

Thistle assured Kay that she and Vincent would be happy with private accommodation, but the Tasmanian people were apprehensive and a bit overawed and booked them into a motel. But Kay thought it over and decided to ask Thistle to stay with her and her family, believing her to be the mother of David Stead's seven children, and so able to cope with three small boys. When Thistle and Vincent arrived by air on July 3, 1967, Thistle did go to stay with Kay, and fitted in well. Kay found out during the visit that Thistle was not the mother of seven children; years later she told Thistle she would never have asked her if she'd known, and Thistle said, Just as well you didn't.[18]

Both Thistle and Vincent gave evidence at a Parliamentary enquiry — a Select Committee had been set up by the Legislative Council to receive evidence concerning the Gordon scheme. Before the enquiry, however, Alan Gray, Kay Geeves, Ron Brown (MLC and chairperson of the South-West Committee), Thistle and Vincent flew in a light plane over Lake Pedder. Before Thistle and Vincent left Sydney, Thistle wrote to Kay saying, "any information you can send us — or suggestions of procedure for the enquiry — will be most welcome as we both feel a little incompetent here but will certainly push the case as far as possible . . . we both intend to use the plane journey to have some domestic WL [Wild Life Preservation Society] talks that we didn't have an opportunity to indulge in here."[19]

Returning home, Thistle wrote to Kay to say how happy she was that the Tasmanians seemed pleased with the result of their visit, but she confessed that the highlight of the adventure for her was the consolidation of her friendship with Kay, and the opportunity to meet her family. "Thank you for all you have done for a cause which I think is so worthwhile and thank you also for your many personal kindnesses. I shan't forget the first week in July 1967, quickly."[20]

Thistle wrote to Ron Brown in July suggesting he seek the support of the I.U.C.N. (International Union for Conservation of Nature and Natural Resources). She had thought Ron should send a cable, but Vincent Serventy had pointed out that a cable could not set out a case, assuming that the I.U.C.N. people were unfamiliar with the situation. "Quite a number of associates are needed to save me from the folly of my own impetuous actions," she wrote. Vincent suggested, more sanely, that Ron should send an air letter, accompanied by relevant material also by air, asking for their support. The Wild Life Preservation Society was associated with I.U.C.N. and Ron was invited to say that the South-West Committee had their support.[21]

In August 1967, the Select Committee Report, approving the Gordon scheme, was tabled; the Report recommended that there be a limitation of fluctuation of shoreline to limit areas of mud flats; that a new National Park be declared; and that a new controlling authority for Tasmanian National Parks be established. It also recommended that an investigation into the flora and fauna of the area to be flooded be undertaken; also a review of the H.E.C. public relations policy, and the referral of all new power development proposals to a Joint Committee of Parliament were recommended. The wildlife survey was carried out and the South-West National Park formulated in 1968, but some areas of the south-west were excluded.

The "Lake Pedder Action Committee" (L.P.A.C.) was formed in March 1971, but Premier Bethune bluntly refused any reconsideration of the issues at stake, despite changing environmental attitudes and failing industrial attractions. Interest increased on the mainland with L.P.A.C. groups forming. In March 1972 the Liberal Government collapsed and Labor came to power, but conservationists lost the legal challenge. The battle had been lost — in 1975 Lake Pedder was flooded. But a new awareness by the public about the value of the south-west had developed, and in 1983 south-west Tasmania was declared a World Heritage area. This time it was a different story when the government proposed the Gordon below Franklin Dam in 1979. In 1984 the High Court of Australia ruled that this dam was in breach of World Heritage values.

Thistle was on the move again in August 1967, planning to drive to Wirrimbirra, then on to Canberra to talk to an SGAP meeting, and down to Melbourne for a meeting of the Australian Conservation Foundation. She left Wirrimbirra early on the Thursday morning, hoping to have a few hours in the Black Mountain Botanic Gardens before the SGAP meeting. But it was not to be. Here is her description of events. "Driving up a hill just out of Goulburn, a car came racing towards me from the crest, well on the wrong side, trying to overtake a semi trailer, and when he saw me he swung over ever more to the right. I was well over and pulled over as far as possible with the result that we crashed head on. Fortunately no one was seriously hurt. I have a very sore chest, a swollen lip and some minor bruises and feel rather miserable as I had great hopes of the ACF meetings. I did think of going down by air on Friday morning but thought it wise to see a doctor and in any case did not really feel up to it. I sent telegrams from Goulburn."[22]

An accident such as this did not restrain Thistle for long because in September she was off to the south-west of New South Wales, botanizing around West Wyalong and the Cocopara Ranges in the Rankins Springs district, which she described as a botanist's happy hunting ground. She had previously spent two brief periods botanizing in the stretch of country between Goolgowi and West Wyalong, and now planned further explorations. She accepted the kind invitation of Mr and Mrs Alan Cruickshank of Rothdene to make their home her headquarters. "Their understanding and tolerance of my erratic comings and goings and their delightful company made this visit an enduring memory." She was particularly gratified to meet a number of other local people who were concerned about the flora of that district and its survival in a climate where hard and constant labour was the only road to success. She visited several native gardens in the district and later said, "I am returning to Sydney feeling very humble, but grateful to those people in the district who are determined that future generations will have an opportunity of enjoying the bushlands which they themselves have cherished".[23]

Thistle returned home to face a hectic week: it was first a working weekend at Wirrimbirra; Ivy Stead was painting the Watson's Bay house with moderate help from Thistle; she bought a new car, alarmed at the NRMA report on her old one; visitors from Coffs Harbour arrived to be shown local reserves, etc.; Saturday was a Conservation Conference; Sunday was consultation with the architect at Wirrimbirra, and there were lectures in the coming week.

In December 1967 Thistle started work on a revision of her 1953 book, *Australian Plants for the Garden.* It would be a major operation, she wrote to Kay Geeves, since so much had happened in that area since its publication. "I am trying to get lists of plants (native) available in the various States for the public and also information on the adaptability of plants in cultivation. Nurserymen's catalogues would be a great help. Could you put me in touch with nurserymen and growers who might be able to help? Do you have any government nurseries?"[24] In sending back thanks for Kay's response, Thistle wrote, "You are right — it is going to be a major job but I shall learn a great deal even if I never get finished. Wirrimbirra is enough to keep me going for two more lifetimes without anything else".[25]

George Ferguson at Angus and Robertson was not so convinced about a second edition of *Australian Plants for the Garden*, believing it would be in competition with Alec Blombery's *A Guide to Australian Native Plants*. But Thistle wrote to Ferguson in June 1970 refuting his claim, saying that the two books were on an entirely different level and each would appeal for quite different reasons. "Blombery gives a useful general approach to Australian plants and their botanical relationship, with some general notes on methods of cultivation. I would be giving much more specific information on individual plants for the information of growers and designers. Also, Blombery was published in 1967. It will be at least 12 months before I could let you have a completed manuscript and with a complex book of this kind, at least two years with the publishers. This brings its publication six years after Blombery which reduces any danger of competition very considerably, if it exists at all."[26]

In 1967 the New South Wales Government had passed the NSW National Parks and Wildlife Act establishing the National Parks and Wildlife Service. Thistle, together with other conservationists, was critical of the Act well before it was passed in Parliament. For more than ten years conservationists had been pressing for legislation to provide better administration of national parks and wildlife conservation in the State. When, shortly after the elections in 1966 when a Liberal Government was returned under R. Askin, the Honorable T. L. Lewis, Minister for Mines and Lands, expressed his intention of bringing forward such legislation immediately, gratification was felt. When, however, the Minister made public the provisions of the Bill, conservationists were concerned by its weaknesses. In order to present a consolidated case for consideration, the Nature Conservation Council of New South Wales, which had fifty associated member organizations at that time, went thoroughly into the contents of the Bill and arrived at a number of basic suggestions for amendment.

The NCC memo, circulated to all members of State Parliament in late February 1967, said, among other things:

"The Nature Conservation Council of NSW believes that the manner in which the National Parks and Wildlife Bill deals with the Fauna and Wild Flowers Acts affords no benefit to the administration of wildlife conservation. In fact, the Bill damages and dismembers a programme which has proved its worth . . . The Bill makes no provision for a whole range of amendments to the Fauna and Wild Flowers Acts which have been necessary for some time and which, we understand, have been framed by the Fauna Protection Panel. We strongly urge that the Panel be consulted in regard to proposed amendments and the best means of effectively integrating wildlife with national parks."[27] When the National Parks and Wildlife Act was added to the legislation of New South Wales later in the year, conservationists were still not happy about it even though the Nature Conservation Council had withdrawn its opposition to the Bill at the last moment when in the words of its Chairman "substantial improvements had been effected."

Thistle was very critical of Tom Lewis and his policies. She didn't trust him, not for any one particular reason, but for a number of reasons. It was Lewis who brought Sam Weems from overseas to be the first Director of NPWS, the Service being an independent section under the Minister for Lands. Lewis may have been aggressive in supporting the NPWS with money and staff, but he was criticized, in private if not in public, for filling a number of senior positions with people from the Lands Department and Forestry; they set the policy, and set up a bureaucracy which was somewhat ineffective. Thistle believed that Lewis tended to give positions of authority to those who were his "mates". She, along with others, also believed that putting National Parks and Wildlife together in one Act was a disaster for wildlife because the latter assumed a secondary position. People did not in general agree with Lewis who was first and foremost a politician who, although willing, lacked the understanding necessary to direct matters of conservation.

In April 1969, when Tom Lewis was still the Minister for Lands, Thistle wrote to Kay Geeves, "Am enclosing cutting from today's *SMH*. This mad Minister has quite recently refuted quite bluntly any suggestion that Commonwealth aid is needed in conservation and then turns up with this scheme! I really think he is completely rattled with the battering he is getting from all angles — not only conservationists — the *SMH* called him many rude names in an editorial following his rash statement that the States did very well in conservation without Commonwealth aid."[28] Mr Lewis' scheme was to declare six areas of Sydney Harbour foreshore land as a single national park, but it involved the relinquishment of some land by the Commonwealth. The Commonwealth land in question involved areas with military installations. The *Sydney Morning Herald* editorial said, "Mr Lewis says blandly that he does not expect any opposition to his proposal from the Federal Government. But it is hard to avoid the suspicion that he is trying to bluff his way through this obstacle".[29]

It was a quick trip to Kosciuszko in January 1968, only just over a week, accompanied by Alison Ashby the wildflower painter from South Australia. Unfortunately she found practically nothing in flower, a very short and dry season, but her main reason for the trip was to talk with Neville Gare, the Superintendent, about the Ranger-Naturalist position. She was also hoping to visit a woman of whom she had heard, cultivating alpines at about the 5,500 foot level.[30]

August 1968, and Thistle was writing to Kay Geeves to make arrangements to meet Kay and her husband Dudley in Melbourne in September; Thistle was anxious for Kay to attend a SGAP meeting and also visit some of her grower friends such as Ivo Hammet. "Off to Coff's Harbour for a week tomorrow," she concluded her letter.[31] She returned from her Coff's Harbour trip with mixed feelings, having gone there to interview the manager of Australian Paper Mills, which had control of extensive areas of land. "The manager was hopping mad at the thought of our visit, which was published locally, and was rather taken aback at our attitude." But she was pleased to see the revegetation on dunes after mining on a coastal strip near Port Macquarie. "They have used native species grown from local seed by a really good nurseryman — a fellow to work on."[32]

Thistle wrote to Ivo Hammet in Victoria to say she would be in Melbourne on the evening of Friday, September 27. "Will the bed still be vacant?" She asked Ivo to arrange visits to as many gardens as possible, including those of Bill Molyneaux, Peg McAllister, Bert Hargrave and Maranoa Gardens. She also wanted Kay to have discussions with SGAP people concerning her plans to start a group in Tasmania. And yes, they wanted to visit the Whipstick mallee. "Would like to cram in as much as possible."[33] For Kay it was a whirlwind tour.

A matter of great concern to Thistle and her fellow conservationists surfaced early in 1968 when mining leases were granted to a cement company to mine limestone in the vicinity of the Colong Caves Reserve in New South Wales. The company was Associated Portland Cement Manufacturers (Aust.) Ltd., A.P.C.M., which was three quarters owned by British investors. On May 29, 1968, representatives of fifty New South Wales societies met at Sydney University and voted Kanangra/Boyd/Colong "the most important wilderness area in New South Wales". The meeting set up "The Colong Committee" to spearhead the conservation movement's fight to save the Colong Caves Reserve from mining. The Premier, R. W. Askin, was asked to take immediate and effective steps to revoke mining licences in and adjacent to the Colong Caves Reserve, and to incorporate the Reserve within the proposed boundaries of Kanangra/Boyd National Park.

In September 1968 the Save Colong Bulletin claimed that the Minister for Mines, Mr Fife, and his cabinet colleagues were consistently stating that Colong limestone caves would not be affected by limestone mining in the Colong Caves Reserve. But the Colong Committee claimed that in the northern end of the Reserve there were three cave systems, parts of which were lying under the existing leases, and the remainder was included in current lease applications.

Thistle wrote to Kay Geeves in May 1969 to tell her that a number of young conservationists had bought shares in the mining companies and attended a shareholders meeting — good publicity, and a promise from the companies to investigate other sources of limestone. In June she wrote that the shareholders' meeting on Colong was fantastic and received quite a bit of publicity. "I have had to withdraw from active participation in the [Colong] Committee because of time."

A year later, in September 1969, Mr Fife was claiming the government could only rescind the leases at the request of A.P.C.M. The Colong Bulletin claimed that the Minister's attitude represented an abdication of his authority and responsibility. "The Government must decide whether it or private commercial interests are to control the affairs of this State." The Colong Bulletin also reported that the A.P.C.M. had stated that it was prepared to consider alternatives to mining Colong. Thistle wrote to Kay Geeves in October 1969, to report that things were uncomfortably moving (that is, uncomfortably for the authorities). "The Minister for Mines has stated that he will investigate alternate limestone deposits and if they prove economic will cancel the Colong leases. This we have been asking for for several years. Ivor Wyatt and I interviewed the Minister for Lands last week about some leases adjoining Wirrimbirra. It was reasonably satisfactory but when we were about to leave he burst out into a tirade of complaints about conservationists in general and said some stupid things. I have written them back to him so that he cannot get out of admitting that he said them."[34]

Meanwhile another issue had come to the fore: the plan to clear the Boyd Plateau and develop it as a 15,000 acre pine plantation. The Colong Committee adopted this issue as one of its concerns, and by 1973 had changed its name to cover the extended objectives. It was agreed to form itself into a national wilderness society, and it still functions at the present time as the Colong Foundation for Wilderness.

By 1973 the A.P.C.M. were prepared to surrender their lease in the Colong (Mt Armour) area, provided they obtained a suitable alternative source. However, the firm "successfully conspired with the New South Wales Liberal Government to take over public reserves on the scenic edges of Bungonia Gorge for limestone mining — in return for which they gave up their equally obnoxious leases over a public reserve at Colong Caves".[35] The mining company had been allowed to extend its lease towards the gorge, to the edge, but not down into the gorge. The Colong Committee had asked for the existing mine to go deeper rather than extend on the surface towards the gorge. Today, in the late nineties, much damage has been done by debris rolling down into the gorge.

The *Sydney Morning Herald* announced on Monday, April 14, 1975, that "Conservation groups have won their five-year campaign to prevent the bulldozing of Boyd Plateau. The Minister for Lands and Forests, Mr Morris, announced yesterday that there would be no planting of pines on the plateau. He further stated that there seemed no logical reason now for not including the plateau in the Kanangra-Boyd National Park."

Printed material produced by the Colong Foundation has this answer to the question. What was achieved? "The saving of the southern part of the Kanangra-Boyd Wilderness was a great achievement despite the offsetting loss of Bungonia, which was already badly damaged. It heralded the dawn of a new conservation era. The 'divine right' of mining interests to wreck any landscape had been successfully challenged, even though the Askin Government never retreated from its attitude that reserves may be mined. The damaging publicity which that Government incurred may well deter any future revocations of reserves for mining purposes. The campaign may be compared to the Lake Pedder campaign, which, though it failed, aroused public awareness and provided the groundwork for the later successful Franklin preservation."

The years 1968–69 were truly hectic ones for Thistle as she persevered with her course at University and worked at

Wirrimbirra. She wrote to Kay Geeves in October 1968, "The Nature Conservation Council is setting up an ad hoc committee to take action on the Sim Report (which is not available except in a few copies) and I did think for an hour or so quite seriously of giving up my University work so that I could do something. But I suppose that would be foolish at this stage".[36]

The Sim Committee had been set up in 1965 by the New South Wales Government with the objective of inquiring into and resolving conflicts between the interests of conservation and the beach mining industry on the north coast. The Chairman was Mr Ian Sim, Town Planner with the State Planning Authority of New South Wales, and Allen Strom was a member of the Committee in his capacity as Chairman of the Fauna Protection Panel and Chief Guardian of Fauna. The report was submitted in January 1968 to Mr Wal Fife, Minister for Mines, and to Mr T. L. Lewis, Minister for Lands. Some positive recommendations resulted from the inquiry: that certain north coast areas be set aside as parks and reserves; that certain parts of these areas be kept in a condition generally free from mining or other disturbance for scientific study purposes; that mining be allowed in certain parts of these areas and not in others; that certain administrative procedures, involving co-operation between the Department of Mines and the National Parks and Wildlife Service, be adopted so that the areas could be effectively used, at the one time, for both mining and nature conservation; that effective restoration work be carried out after beach mining.[37]

Thistle added in her October letter above that the Professor of Civic Design at New South Wales University, with whose students the Landscape Design class was taking a seminar, had invited Tom Lewis to talk to the class. "Won't he be pleased to see me? We were invited to send him questions in advance — mine was, would he be prepared to release copies of the Sim report to the class for study? The Prof. was shocked but I think he did it. The Committee was heavily loaded by Mining interests. The report will probably be tabled in Parliament which was not the Minister's first intention."[38] In her first year of the course, Diploma in Landscape Design, Thistle submitted an assignment of nine foolscap pages on "The Origin of Cultivated Plants", for which she received six and a half out of ten. The lecturer's comments were, "This is a mine of information on early cultivated plants. I appreciate the effort which has gone into its compilation, and the direct language of the accompanying text, but it seems a shame that a little more time was not spent in presentation, i.e., in designing an attractive format and in producing a faultless piece of typing. I know that time was against you all, but the value of presentation must not be underestimated."[39]

Thistle was always appreciative of the help she received from her fellow students and the staff during her Diploma course. "Most of them were architects of several years' standing and they regarded me as a kind of mascot. They were benevolent to me but at the same time they demanded very high standards . . . they were highly critical but constructive in criticism . . . I would like to say one small thing in a general way about university students and my experience with them on the campus of the University of New South Wales, because I was surprised. I thought, 'Here's a woman coming along, not in the first flush of youth, they'll probably be extremely contemptuous of me'. I found them not only tolerant but extremely courteous, and not only courteous but tremendously interested in what they were doing. One of my delights was to go down to the Union and just sit around and watch them and listen to them. I must say they are not very elegant to look at in the way they dress but this is not very important after all. I did find that their conversation was intelligent, that their topics of conversation were relevant to their studies and my admiration and respect for young people was increased very much by my two years spent at that university."[40]

Into her second year of study she was finding the course "very heavy going". "The design section is very difficult for me as my aptitude for this is well below par and although the boys do their best for me I am truly well behind. But I have indeed gained a great deal in other ways through the course and the class as a whole is becoming very conservation minded. Perhaps

I should repeat with the next group!"[41] In October things were really serious. "If you don't hear from me for a while it will be because I am in gaol. I am one of the 38 who signed the Act of Defiance on the National Services Act and have been found guilty and fined 50 dollars. This of course will not be paid (time up yesterday) and today it was announced that the Attorney General said he would issue warrants. So I may be sitting for the exams in solitary isolation (it is all arranged). You can understand I am a bit busy tidying things up — time is 25 days. I hope some off for good behaviour — will try."[42]

"Exams over and no warrants yet — the magistrate was requested on Friday to issue these and refused. Now it goes to the High Court (Monday) but the issue has become more complicated since there are now another 30 added to the list of 37 already fined and another 50 to come (probably before the same magistrate) this week."[43]

"It has been impossible — so far — to get into prison but we have scored some good publicity. The magistrate refused to prosecute another group today and we grow stronger every day. I was terribly tired when the end of University term came and I had not realized how much had to be done in my own nursery. However, I am now recovered and plotting bigger and better things for 1970. I think we might get Wirrimbirra into the nursery business in quite a big way and we have a big education programme to organize down there too. WLPSA and some others are planning a Wild Life Show in Sydney next August so it is just as well no more University courses are coming up."[44]

One of Thistle's fellow students in the Diploma in Landscape Design was Dr Sid Baggs, who did his Master's and Doctorate after the Diploma course. Sid remembers that when she was a student in Professor Peter Spooner's classes, quite a few of her fellow students, professional people, found her abrasive. Sid got on well with her, he remembers, because he liked a forthright woman who said what she thought — it appealed to him. She seemed to have a "yang" or "butch" side to her personality. Sid held a lecturing position at the University of New South Wales in the School of Landscape Architecture from 1975 to 1986. Some student projects in his subject, Environmental Impact Assessment and Evaluation, involved a botanical input and in the early eighties, Sid invited Thistle several times to lecture to his students. She was still very sharp mentally, he recalls. Sid and Thistle also worked together on a low key project for the Woollahra Council at South Head in the early seventies; Thistle supplied the major input, with consultative advice from Sid.[45]

At a much later date, 1988, Thistle wrote a short piece entitled, "My Interest in Landscape Design", and her personal letterhead said, Thistle Y. Stead, Landscape Architect. After discussing briefly her early interest in botany and the founding of Wirrimbirra, she comments on the positive outcomes of her course at the University of New South Wales. "As well as assisting materially in the design of Wirrimbirra, it had a profound effect on me. It altered my own relationship to the bushlands through which I passed. Prior to this, I would comment to myself or trapped passengers on the species likely to be found in particular habitats. But, after a little tuition, my imagination led me to consider the detailed natural landscape as I drove past, and to ignore the species comprising it, giving greater attention to the combination of species and their relationship to one another, and why. I found this a most fascinating study (and a great detriment to speeding) . . .

"The more important result, personally, was a closer understanding — while some areas emphasized certain particulars, others, apparently similarly situated to them, presented others. This was extremely important to me in considering design. We needed to consider the relationship of the areas to be treated with the rest of the property so that there would appear to be continuity with the areas available for planting, the inevitable losses on land required for building, etc. For many reasons, chiefly the lack of personnel, our plans were not entirely successful."[46]

In 1969, Christina Stead had made a brief visit to Australia. She was now a widow since the death of her husband, Bill Blake, in London in February 1968. The Australian National

University in Canberra had awarded her a Creative Arts Fellowship, a short-term B class fellowship: a category for Australian expatriate artists of international repute. The lump-sum stipend, $5,000, was to cover her fare and expenses. The letter from the university stated that there was no obligation to give lectures or talks, but it was hoped that Fellows would meet and talk, at least informally, with members of the academic staff and students. Stead would be given an office in the English Department.[47]

Christina left for Australia on the evening of July 28, 1969; she had sailed for England at the age of 26, a budding writer. Now she was 67, a writer with an international reputation. But once installed in Canberra her role as a Fellow made her uneasy. Unused to being treated as a celebrity, she encountered it at a time when she had neither Bill nor current writing projects to make her feel good about herself. She worried that she was not carrying out her obligations satisfactorily. Her few trips to Sydney were not a success; she avoided Thistle and never once visited the house at Watson's Bay.[48] However, she did spend a pleasant fortnight with Gilbert Stead and his family in a holiday house in Mollymook, on the coast south of Sydney. With her brother David and his wife Doris, she drove west, towards Broken Hill, and found the sunburnt country a thrilling experience. But by mid November Christina was flat and exhausted and anxious to be home — Australia was not home. She left on December 3, 1969.[49]

Thistle's course in Landscape design was successfully completed, her neglect of Wirrimbirra and her Alpine book were no longer on her conscience, and she looked forward to 1970 with her usual enthusiasm and commitment.

Chapter 7 Notes

1. *Australian Wild Life*, Vol. 4, No. 5, June 1965.
2. Vincent Serventy, pers. comm. 19.10.96.
3. ibid.

4. Eve Gourlay, pers. comm. 18.2.97.
5. K. Geeves, pers. comm. 11.1.97.
6. K. Geeves, pers. comm. 20.2.97.
7. T. Y. Stead, "Island of Enchantment" in *Australian Wild Life*, Vol. 4, No. 5, June 1965.
8. ibid.
9. ibid.
10. ibid.
11. A. Sourry, pers. comm. 16.11.96 and 18.5.97.
12. ibid.
13. ibid.
14. Reported in a local newspaper, June 28, 1967, cutting in the possession of A. Sourry.
15. Copy of letter supplied by K. Geeves.
16. Letter from T. Stead to K. Geeves, 21.6.67.
17. ibid.
18. K. Geeves, pers. comm. 11.1.97.
19. Letter from T. Stead to K. Geeves, 23.6.67.
20. ibid. 10.7.67.
21. Letter from T. Stead to R. Brown, 19.7.67.
22. Letter from T. Stead to K. Geeves, August 1967.
23. *The West Wyalong Advocate*, October 3, 1967.
24. Letter from T. Stead to K. Geeves, 30.12.67.
25. ibid. 29.1.68.
26. Angus and Robertson correspondence, ML MSS 3269 Vol. 324.
27. Information from an Editorial written by T. Y. Stead in *Wildlife Research News*, March 1967.
28. Letter from T. Stead to K. Geeves, 14.4.69.
29. *Sydney Morning Herald*, 14.4.69, Editorial, page 2.
30. Letter from T. Stead to K. Geeves, 29.1.68.

31. ibid. 20.8.68.
32. ibid. 29.8.68.
33. Letter from T. Stead to Ivo Hammet, 10.8.68.
34. Letter from T. Stead to K. Geeves, 30.10.69.
35. "Save Colong Bulletin", May 1974.
36. Letter from T. Stead to K. Geeves, 18.10.68.
37. Report of the Committee of Enquiry on Differences and Conflicts Between Interests of Parks and Conservation Authorities, Scientific Bodies and Mining Companies, Ian McH. Sim, January 1968.
38. Letter from T. Stead to K. Geeves, 18.10.68.
39. ML MSS 5715, 4/25.
40. de Berg Tape, Side 2, pp. 7–8.
41. Letter from T. Stead to K. Geeves, 19.9.69.
42. ibid. 30.10.69.
43. ibid. 23.11.69.
44. ibid. 10.12.69.
45. Dr Sid Baggs, pers. comm. 13.2.97.
46. ML MSS 5715, 2/25
47. H. Rowley, p. 483.
48. ibid. pp. 488, 489.
49. ibid. p. 490.

CHAPTER 9

A.M. — For Wildlife Conservation — 1970–1980

"The conservationist is one of the most important people living, because civilized living is not civilized, properly speaking, without the natural environment of which man himself is a part."

T. Y. Stead

THE year 1970 was a busy one for Thistle at Wirrimbirra, but she also found time to prepare evidence to be presented to the House of Representatives Select Committee on Wildlife Conservation. On May 14, 1970 the Select Committee had been formed by the Federal Liberal Government, John Gorton Prime Minister, under the chairmanship of Mr E. M. C. Fox. The report on the Committee's findings, released in October 1972, was unofficially known as the Fox Report. The Committee was in Sydney in December 1970 to take evidence. Thistle attended twice, once as a representative of the Stead Foundation, supporting Allen Strom, and once on behalf of the Wild Life Preservation Society of Australia, supporting Vincent Serventy. Thistle herself was to make a submission at a later date, Wednesday, February 2, 1972. Allen Strom was called upon to speak to the submission made to the Committee, in his capacity as Chairman of the Board of the Stead Foundation. He stressed two aspects of wildlife conservation as seen in the work of the foundation: firstly, efforts of private citizens in preserving naturalness as a separate but extremely important contribution beyond the national parks and nature reserves; and secondly, the paramount significance of conservation education so well demonstrated in the work of the Foundation at Wirrimbirra.[1]

Vincent Serventy appeared before the Select Committee on December 4, 1970 in his capacity as President of the Wild Life Preservation Society of Australia. Serventy presented figures to illustrate the position of national parks and fauna reserves in Australia, claiming that the low figures (1.3% in New South Wales, 0.6% in Queensland, 1.1% in Victoria), compared with the five per cent regarded by experts as a minimum, was a shocking indictment of governments, and constant pressure was needed to improve the situation. He then made comments on the major reserves to clarify his terms of reference — national parks, nature reserves, wilderness areas, anthropological reserves. Eight points were put forward as a basis of his recommendations — adequate classification of the environment, monitoring of the Australian environment, income tax incentives for preserving, rather than clearing as at present, a Federal National Parks Authority, uniform conservation laws, more research, training of personnel, and more research on nuclear power. Ensuing discussion with members of the Committee ranged far and wide, and Serventy was asked to clarify certain proposals he had made. Eventually (after 86 pages of transcript) the Chairman accepted the submission.[2]

Thistle's submission as Vice-President of both the Stead Foundation and the Wild Life Preservation Society in February 1972 was supplementary to those presented by Allen Strom and Vincent Serventy. She opened her presentation by saying that she wanted to cover four fields in which she felt that the role of the Commonwealth Government should be greatly strengthened. These were:

1. Assessment of environmental deterioration and necessary action to mitigate this.
2. Legislation concerned with wildlife conservation.
3. The role of voluntary organizations in wildlife conservation.
4. The education programme in wildlife conservation.

Thistle spoke on each of these aspects and made eight categorical recommendations: four recommendations referred to environmental deterioration (1–4); recommendation 5 referred to legislation; recommendation 6 referred to the role of voluntary organizations; and recommendations 7–8 referred to education. Two of these recommendations are given as examples of her position.

Recommendation 1: That the Commonwealth Government undertake, in co-operation with the individual states, a survey of land use potential throughout the continent with recommendations for future usage, including the provision of national parks, nature reserves and recreational areas.

Recommendation 7: That the Commonwealth Government initiate a research programme in conservation education techniques and processes at primary, secondary and tertiary levels.[3]

The Select Committee set up under Fox was probably the first time the Federal Government had become involved in environmental matters in an attempt to find out the needs of wildlife conservation. It would appear that the recommendations of the Fox report were unable to be addressed because there was a Federal election in 1972 and the Labor Government took over from McMahon's Liberal Government. However, it is more than likely, as Vincent Serventy suggests, that the work of the Fox Committee contributed to the deliberations of the Hope Committee, set up in May 1973 by the Whitlam Government to look into the nature and state of the National Estate. The Committee, under the chairmanship of Robert M. Hope, released its report in 1974, a most comprehensive report, addressing both the natural and built environment, Aboriginal sites, and cultural property such as museums and archives. In addition to discussing the role and responsibility of the Australian Government, the Committee addressed Education for Conservation, and the role of Voluntary Organizations. Those who held discussions with the Committee included the Wild Life Preservation Society of Australia, the Nature Conservation Council of New South Wales, the National Trust, and the David G. Stead Memorial Wildlife Research Foundation of Australia.[4]

By the sixties and seventies Thistle had developed her skills in photography. "Most of my photography I like to do in the field, I like natural light on them and I like to get the natural appearance of the whole bush. Sometimes I get myself into great predicaments over this. I was down in National Park once, lying flat on my tummy with a tripod beside me, taking a photograph, and a motorist pulled up, with great shrieks of brakes, and a young man got out and started rushing towards me, so I just waved the tripod at him to let him know that all was well, and off he went.[5] This outdoor photography does involve a lot of time, particularly in alpine areas where the light is so precarious, and you set up your tripod, get all ready, take a light reading, just about got to press the trigger and you find that the light has changed entirely, so you get another half hour's wait. I have waited several hours just to get one photograph. I like to photograph alone. I find particularly photographic friends in the field most embarrassing . . . the other sort of people that I don't like are people who want to hold things for me, because this is very confusing, I like to be entirely alone."[6]

"**Noted Conservationist and Biologist Visits Tasmania**," was the headline for a newspaper article on Thistle in Tasmania in February 1971. As early as July 1970, Thistle had been making arrangements with Kay Geeves for a visit in January/February 1971. The Regional Officer of the Adult Education Board in Burnie, Tasmania, Mr A. McG. Campbell, wrote to Thistle in September to invite her to conduct a weekend school at Cradle Mountain during her proposed visit to the area. She answered him to say she was willing although admitting she didn't know much about the region, advising him to consult with Kay Geeves about arrangements.

Thistle travelled to Burnie with her car on the *Empress Of Australia,* arriving on Friday, January 15, 1971, a short drive to Launceston where she stayed that night. The next morning she drove to Cradle Mountain to meet the group, most of them members of the Burnie Field Naturalists' Society. Kay Geeves had driven up from Hobart with her husband, Dudley, and three young sons; also with them was Melva Truchanas, a bushwalker of many years standing and a keen student of native flora. On Saturday night Thistle gave a slide presentation on alpine flora; Saturday afternoon and Sunday morning were taken up with field work. Melva (whose husband Olegas was to die tragically in an accident in January 1972) had become the inaugural Secretary of the fledgling Tasmanian SGAP group; she recalls her first meeting with Thistle at Cradle Mountain. "She lived up to the walking mythology which Kay had built up for her — the guru of Australian flora. The inspiration Kay derived from Thistle was handed on to the Tasmanian people. My first memories are of a day field trip at Cradle Mountain, showery weather, an early autumn. She was knowledgeable, greatly enthusiastic and curious — after all, she didn't know all the Tasmanian plants.

She was brilliant at social interaction, as long as you were interested in the same thing — she did not tolerate fools gladly."[7]

On Sunday Dudley Geeves returned to Hobart and home at Port Huon, but Kay and the boys stayed a few days extra with Thistle at Cradle Mountain. On the following Friday, January 22, they met a friend of Thistle's, Doris Davidson, at the railway station at Deloraine, and the group travelled south to Port Huon where Thistle and Doris stayed in an apple pickers cabin on Kay's father's property.

About thirty members of the Tasmanian group of the SGAP attended a gathering at the home of Miss Essie Huxley of Longley on Wednesday, January 27. Members travelled from as far away as Burnie and Launceston to hear Thistle Stead speak on "Grevilleas for the Garden". Then it was a trip to Mt Field National Park for six days (February 1–6) until Doris had to leave on the 7th and Thistle met her friend, Margaret Senior, at the airport the same day.

The Hobart *Mercury* reported on February 17, Women's Page, "Representatives of many Hobart garden clubs were among the capacity audience at the meeting of the Society for Growing Australian Plants, Tasmanian Group, at the Country Women's Rest Rooms, Hobart, on Monday night — February 15). Miss

Harris, who illustrated her talk, spoke about mountain flora in Tasmania, and alpine plants native to Australia. Covered buttons and loops trimmed the white linen frock worn by Miss Harris. At the V-neckline she tucked in a red and white spotted silk scarf." Thistle wrote to Kay after she arrived home, "Thanks for the cutting — so pleased to know what I wore".

Margaret Senior and Thistle left Hobart on February 16, the car stacked with plants and cuttings. They stayed in Launceston, to visit friends, including Mr and Mrs Herbert J. King, Mr King a greatly respected member of the Launceston Field Naturalists' Society. Then to Burnie to visit some nurseries and native gardens started by local enthusiasts, and on to the ship. A letter to Kay soon after arriving home expressed her satisfaction about the trip. "I had grand plans of spending the voyage on a long letter to you and some other letters as well as running through 'Alpine Plants' for errors. But a following wind and choppy seas settled that and although I was not exactly speaking seasick, I managed to stave it off only by lying prone most of the time. So I had that enforced rest after all . . . My car was the very last one off the *Empress* at Balmain and I had to wait one and a half hours which did not please me. However, I managed to get all the plants and cuttings unpacked and under mist that day. The plants travelled surprisingly well . . . Things seem to have stood still during my absence (how I do flatter myself) but are beginning to pile up — ACF Education Committee tomorrow, WLPSA annual meeting Tuesday, Field Studies meeting Wednesday, Allen Strom has a school for me to see at Gosford pronto, my colleague architect is waiting to start the Mably school we are to do together. Wirrimbirra will have to wait a few days — not very much rain in January but better in February . . . Be reassured, I shall have many early nights (commencing tonight) and shall be as fit as you say you are in no time. Pat Stanley, my dear Canisius College brother writes — 'It's more important to conserve Thistle Harris than any of the plants,' with which you would undoubtedly agree."[8]

It was now June 1972 and Thistle was approaching her seventieth birthday. "On the eve of the Concorde visit —

rushing out to Mascot early morning to register my protest. A meeting earlier this week and 100,000 people are expected to be there. On Sunday there is a protest meeting in Moore Park against the proposed sporting centre. Life is full of protests. So glad you are out of apples and sorry to have been so long in writing — no apples for me but lots of other things."[9]

About mid-year 1972 Thistle also started a series of lectures on wild flower identification for members of the National Parks Association. Fifty in the class, and twice too many, she said, but very enthusiastic. They were much overbooked, and it was Thistle who insisted on closing the doors at fifty participants. Six lectures and two field days completed the course.

In the years 1968–1974, Thistle formed a close friendship with Elizabeth (Betty) Chandler, who during those years was a great support to the work at Wirrimbirra. Elizabeth was particularly interested in the study and cultivation of Acacias, and it was actually through a mutual acquaintance who held the SGAP seed bank of Acacias that the two enthusiasts were to meet. Thistle told Elizabeth that the Acacia study was too large a concern for her to deal with because of all her other interests, and she asked Elizabeth to be the convenor of the Acacia Study Group at Wirrimbirra. She was to be known at Wirrimbirra as 'Wattle Betty.' She also served as Assistant Secretary to Kathleen Groat's position of Secretary during those years, and was often Acting Secretary. The close contact Elizabeth had with Wirrimbirra ceased after a serious car accident in 1974. "We did a lot of reckless things together — went interstate every year, several times to Melbourne, and to Queensland, and to Kosciusko."[10]

Sometime in the early seventies, Thistle and Elizabeth travelled to Melbourne with John Teulon. "Wonderful! John was driving!" They travelled down the coast road; when Thistle saw a plant she wished to examine, it was Stop! and John hardly had time to brake the vehicle before Thistle had the door open and her feet on the ground. They stayed the night at Maffra in Victoria to visit the nursery of Bill Cane, then on to Melbourne.

It was November, and as they neared the capital city, Elizabeth said, "Do you know what day it is? No? Well, it's Melbourne Cup Day". To this information there was no reaction. So what? Thistle asked. No accommodation. Thistle never booked accommodation ahead when she travelled. On their arrival day they had arranged to have dinner with Lady Doreen Dew and her daughter Libby; Lady Doreen obliged by ringing around to find accommodation — the result? They stayed at the Dew residence.[11]

John Teulon was production manager at Swane's Nursery in Dural. He had met Thistle at the nursery about 1967 when the cultivation of native plants was becoming more widespread. She was to be a great help to the nursery over the years with her advice and botanical knowledge. But during the early seventies when she was a frequent visitor her visits were not always appreciated, because instead of leaving her car in the carpark, she would drive around the nursery, knocking into their water pipes, often backing into them, and usually causing a great deal of damage. Of course she was contrite, but the maintenance man would go berserk. After five or six episodes, Ben Swane said, "Enough. You can't drive around the nursery any more. Leave your car in the carpark!"[12]

Thistle had contacts with the nursery people all over the country, but was not always in sympathy with the commercial side of the industry. She was more interested in growing the rare and the unusual than in what would bring in the dollar. John Teulon tried to explain to her that his concern was bringing on plants that people would buy. She was a frequent visitor to Sid Cadwell's nursery Bongala at Annangrove, but probably wore out her welcome because Sid was a recluse and didn't encourage people to call. On one such visit with a friend, Thistle had taken food for a picnic lunch, but had forgotten to pack a spoon. She asked Sid if she could borrow one. After some hesitation he said that would be pretty difficult as they were all in the slug defender. Sid moved to Rylstone in 1975 but Thistle thought of him as an old friend and wanted to visit. John Teulon saw Sid frequently but refrained from telling Thistle. She said to him one day, "If you go to see Sid and don't take me, I'll cut your throat". When John told Sid, the latter said, "If you do bring her here, I'll cut your throat".[13]

Thistle said the trouble with the nursery people was that they fell into two categories: there were the very successful commercial ones who were out to make money, and then there were the absolute enthusiasts who didn't care about money and just wanted to grow the plants. One enthusiast she visited frequently was George Althofer and his brother Peter at Nindethana, Dripstone, near Wellington, New South Wales. George had begun growing native plants in his garden in 1932, establishing his Nindethana nursery in 1938 as a local industry, under primitive conditions, using only dam and tank water. It grew quickly, and from the 1940s he found an expanded market, introducing many species and sending plants Australia-wide, as well as to New Zealand, the United States, England, Ireland, and to countries in Europe.[14] Peter and George were the prime movers in the establishment of the Burrendong Arboretum. Many of the nurserymen Thistle befriended were a great support to her in the work at Wirrimbirra. Gifts of plants, seeds, or cuttings arrived regularly from Victorians N. and E. Cleeves, W. Molyneux, A. Hargrave, W. Cane and I. Hammet, as well as from more local growers like Swane's Nursery and Sid Cadwell. Of course right throughout the seventies such gifts for the Wirrimbirra nursery arrived also from individual enthusiasts. George Althofer wrote several books, and in one of them, *Cradle of the Incense — The Story of Australian Prostanthera* (1978) he included Thistle among the Acknowledgements — "To Mrs T. Y. Stead for help over the years with specimens, plants and information that is of more value than gold".

Another man whose work Thistle admired very much was that pioneer of Australian landscape design, Ellis Stones (1895–1975). Ellis's artistic ability with stone in the landscape was without peer. Thistle described Ellis as "that superb worker in rock," writing, "Rock and boulder outcrops demand the highest of skills because of their dominant nature: errors and deficiencies will be obvious because lavish planting cannot be allowed to camouflage them; . . . Few people have the innate ability to use rock in the superb manner (Ellis Stones) did." *The Ellis Stones Garden Book*, an illustrated compilation of Ellis's *Australian Home Beautiful* articles on which he had been working for some time with Tony Hitchin, appeared in 1976, its foreword written by Thistle who concluded, "I really cannot do justice to the work of a man I admired so much, but I appreciate being given the opportunity to try".[15]

In September 1972 Elizabeth and Thistle travelled north to Queensland, the main objective being to visit Dave Gordon at Glenmorgan. The route north was along the coast, with a stop at Lake Cathie to inspect the rehabilitation after sand mining by the Clutha Development Company. Thistle was not impressed by what she saw. When they reached Southport, she went to the offices of the Clutha Development Company and insisted on seeing the manager, somewhat mollified when that gentleman displayed a degree of sensitivity to the damage they had caused.[16]

Dave Gordon and his family had a large pastoral property about 112 km west of Toowoomba, but his passion was for native flora, and at various times he even employed a curator for his native garden. He had plans at this time to make available 10–20,000 acres for an Environmental Studies Centre, remodelling his shearing quarters for student accommodation. The popular hybrid of *Grevillea*, Robyn Gordon, was first grown in the garden of David Gordon, and was named after his daughter. At the time of Thistle's visit in September 1972, a long period of drought had placed the Gordon family in a trying situation; Thistle arrived with her car stocked with provisions to share with these friends, but left them with a greater gift — it started to rain.[17]

Kay Geeves spent a week with Thistle in late August or September 1973, on her way to the SGAP Federal Conference in Brisbane. Thistle was keen to show her the highlights — "Wirrimbirra is a must . . . anyway give me your ideas and I'll tire you out". They did visit Wirrimbirra, as well as the Waterboard country nearby to see the white waratah, and on to Canberra the next day. Thistle would have been delighted to enrich the friendship she shared with Kay; in a letter she had written sometime prior to this Thistle had said, "Friends I have and very good ones, as you well know, but the quality which makes for the sort of companionship we have is, as you are well aware, only too rare. I can only regret that David did not live to know you too for he would have enjoyed you (and teased you too) and appreciated what you meant to me."[18]

Only a month later and our indefatigable traveller was on the road again, Elizabeth Chandler her companion once again. The first stop was at Binna Burra in the Lamington National Park, Queensland, where Thistle was a guest lecturer for three days at one of the general interest courses the management organizes for guests. She was one of several lecturers, and she spoke on "Growing Native Plants", her enthusiasm and presentation endearing her to the patrons. They were anxious to hear something about Wirrimbirra, so she said she would give an extra talk one night after supper. The lecturer scheduled for the talk that evening before supper was an SGAP man, and an engineer, and he spoke a little too long on the drainage of soils, a topic the listeners did not find interesting, and they became restless as they waited for supper and Thistle's talk. Although it was late, Thistle's talk after supper was well attended and well received.[19] Thistle reported to Kay Geeves, "We had a fine time at Binna Burra. Lovely country and a good group of people. Left with our copy book unblotted and made a triumphal exit to a fanfare of trumpets."[20]

From Binna Burra they made their way to Toowoomba, where they completed Thistle's collection of plants. But the timing was bad — it was the city's floral festival and Thistle had not booked ahead for accommodation. After some searching they were fortunate enough to find a vacancy in a motel. South to Goondiwindi where the border patrol was not strict about plant inspection, and down to Mt Kaputar National Park. They arrived at sunset, seeking accommodation in one of the lodges there, but these were booked out. There was no choice but to unload all the plants in their pots from the station wagon and sleep in the vehicle — a cold night and no food. The next morning, not far out from Mt Kaputar and the vehicle had a flat tyre. Thistle had never changed a tyre, but she took out the jack and placed it down upside down, saying to Elizabeth that no man who came along could resist stopping to tell a woman she had it wrong, and would fix it up. Along came the postman — he stopped and changed the wheel.[21]

The Warrumbungles was the next port of call, with the plan being to stay in one of the old tram cars, but these too were booked out. "Thistle should have known but never thought ahead." They sat beside the car at sunset feeling depressed as they ate their take-away food, but caught sight of a car driving out. The office was closed, but Thistle persuaded the reluctant Elizabeth to call upon the ranger, who listened patiently while Elizabeth explained that it was Thistle Harris who needed a bed, and that they had already spent one night in the car. Without a word he handed her the key, and next morning when Thistle entered the office to pay, the ranger prevailed upon her to sign several of her books which he possessed.[22]

A letter to Kay Geeves in November 1973 reported, "Our last expedition was fabulous, full of little miracles, one each day. One small miracle will interest you. Our fan belt broke on the lonely road with a storm coming up. I had a spare with me (don't usually carry one but John[23] had given this to me for some reason not specified). We drove over to a nearby farmhouse (there are not many on this road so close to the roadway) to meet a truck just leaving the farm to bring in a bull on account of the coming storm. Being in a hurry they whipped the belt off in no time and sent us into the missus to get some water for the radiator. She kindly filled it and gave us a bottle to carry water which was most useful for plants — miracle in three parts."[24]

It was probably in December of this year, 1973, that Melva Truchanas from Tasmania took her children to Sydney for a treat, friends having let her stay in their unit at Double Bay. They spent New Year's Eve at Watson's Bay with Thistle, celebrating happily with apple cider, which no one realized was alcoholic cider. Thistle drove them back to Double Bay, the kids on a high, yelling out through the back door of the station wagon to passers-by, "Happy New Year". Melva, reading the true situation, was concerned about Thistle driving back. It was the first of several meetings with Thistle in Sydney for Melva, "always an event".[25]

Meanwhile, on the other side of the world, Christina Stead was struggling with loneliness and poor health. She hated the cold of the English winter, but did not really want to return to Australia to live. Without Bill Blake, the man she had loved, and who had loved her, there was nothing to live for in England, so she made the difficult decision to return to Australia at age 72, after forty-six years in the northern hemisphere. She paid approximately $7,000 for an extension to be built at the back of her brother Gilbert's house in Hurstville and settled in soon after her arrival in August 1974. But she did not settle, she did not feel at home, and openly avoided Thistle, just as she had in 1969. Thistle never knew why. Only once did Christina go back to Boongarre. It was in May 1976, two years after her return. The magazine *Time* was printing an article about her, and their photographer, David Austen, rang Christina to suggest he come out to Hurstville to photograph her. Gilbert Stead was in the throes of renovations and the house was dusty and noisy. Austen suggested he should photograph her in her childhood home at Watson's Bay, and she agreed. He photographed her walking on the jetty in front of the house. Thistle came out and spoke to them, but Christina was distant and ill at ease. The encounter left Thistle baffled and disappointed. [26]

In September 1974 Thistle and Elizabeth Chandler spent a magnificent two weeks in Central Australia. Elizabeth's daughter, Kathy, had recently been married and the young couple wanted to give her a trip to Central Australia as a token of their gratitude for a generous wedding. Elizabeth booked into an organized tour, Desert Trek, two weeks in the centre in four-wheel drive vehicles. Thistle was enthusiastic about the idea of a trip to the Centre, but was decidedly against an organized tour, she wanted to drive her own car with Elizabeth as passenger. At Wirrimbirra one day Elizabeth was discussing her dilemma with Thistle's longtime friend, Dora Reiach. Dora said, "Don't let Thistle talk you out of the trip you want," so Elizabeth stood firm and Thistle agreed to go on the organized trip, and thoroughly enjoyed it.[27]

The two women flew to Adelaide and on to Alice Springs where they met the group, two other women and two men, and two young male drivers in charge of the two 4 wheel drive vehicles. At Alice Springs Elizabeth had arranged a meeting with botanists at the Arid Zone Centre, some 8 km out of town. They were most interested in an arid zone Acacia, an unusual plant which looked like a pine tree. An invitation to spend the evening at the botanists' home was eagerly accepted. Standley Chasm and Palm Springs were visited, then on to the Hermannsburg Mission and Papunya, another Mission Station in the west of the Territory. Thistle was very depressed by the Aboriginal reserves, "I can see no possible hope for the majority of these people who are living between two worlds — in their own humpies for the most part but coming into the stations for handouts when they are available, and having no idea what it is all about".[28]

The vehicles left civilization to travel to the north-west, over into Western Australia, our botanists entranced by the unique vegetation, and back to Mt Olga and Ayers Rock. After a week on the road, the ablution block at Ayers Rock campsite was well used. Then south to Victory Downs, a large pastoral property, and down the Stuart Highway into South Australia. A stop in the Flinders Ranges and south to Adelaide, where they rushed to the showers.[29] Two days were spent by Thistle and Elizabeth discussing with education officers and teachers at the town of Elizabeth a proposal to establish a field studies centre in the Para Wirra Recreational Park. "It was all very invigorating and I hope profitable but I already have a heavy cold — or something and got back to Sydney just about able to crawl. Am still recovering. However, we both still drool over the fortnight away and I have some good photographs."[30] The day after they arrived home, Elizabeth was hit by a car on a pedestrian crossing; her injuries were not serious but her days of excursions with Thistle came to an end.

The Wild Life Preservation Society of Australia continued to have a falling membership into the mid seventies. Vincent Serventy had been President since 1966. In 1975 he spent a year travelling overseas with his wife, Carol, and their family. Thistle became President for the year 1975. When Vincent returned home in January 1976 he received a call from fellow member Dick Mason to alert him to the fact that Thistle had plans to wind up the Society. The Society had some assets and Thistle wanted to merge with the Stead Foundation so that the assets would be available for Wirrimbirra. Dick Mason was extremely unhappy about this — "it was a grass roots society for people in the community interested in the preservation of wildlife, and even though it was not in its heyday, it was a society with a proud history — it should not be abandoned".[31] Dick Mason asked Vincent to attend the AGM in February at which Thistle was going to move the motion to wind up the Society; Dick also asked Vincent to accept nomination to be President once more. Thistle, unaware that Vincent had returned from overseas, was at the AGM supported by her friends, Gwen and Aileen Hanley, and Joan and Eileen Bradley. When Vincent Serventy walked into the hall there was a sudden silence. He was duly nominated, elected President, took the chair and proceeded to open the discussion on the future of the Society by speaking against the proposal. The discussion became heated, and at its height, Eileen Bradley collapsed. Dr Dick Mason tried to revive her without success — she was dead. Vincent had no choice but to close the meeting and adjourn until March, when the matter would be decided. Thistle and her friends did not attend the March meeting, and she never attended a meeting of the Society again, nor took a position on the Council. Vincent was particularly angry with Thistle that she proposed to allocate the funds to Wirrimbirra rather than to the branch of the WLPSA in Queensland. After this setback, Vincent Serventy worked hard to raise the profile of the Society, and succeeded. Today, as a lobby group, it is stronger than ever.[32]

The Bradley sisters, Joan and Eileen, were staunch supporters of conservation and pioneers in bush regeneration after its infestation by weeds. They published *Weeds and Their Control* in 1967, and worked in the bush themselves, around Ashton Park, Mosman, with systematic hand weeding in a way that favoured the native plants. In 1971 Joan Bradley wrote *Bush Regeneration* in which she tried to convey the delight she and her sister felt in helping the bush to help itself. But they were also to experience disappointment and hurt when the brilliance and simplicity of their regeneration techniques were spurned by ignorant and sometimes pigheaded individuals and bureaucracies. The turning point for bush regeneration came in 1975 when the National Trust commissioned Joan, Tony May, and their small band of regenerators to demonstrate their techniques in Blackwood Reserve, Beecroft. With the support and sponsorship of the National Trust and, particularly, the management and promotion skills of Evelyn Hickey, the Trust's Regeneration Officer, the demand by local councils for the services of trained regenerators grew rapidly. Eileen did not live to see this recognition, but Joan was able to provide the skills when a school was established to teach the Bradley method, now being used throughout Australia and in some countries overseas. Joan died in 1983, but a book was published in 1988 based on Joan's manuscripts and edited by three friends and colleagues, *Bringing Back the Bush — The Bradley method of bush regeneration*, edited by Joan Larking, Audrey Lenning and Jean Walker.[33]

Thistle did not allow the matter of the Wild Life Preservation Society to dampen her spirits — she was much too busy with her next series of books, *Gardening With Australian Plants*, the first volume of which was on Shrubs, and was released in 1977. Thistle was assisted in the proof-reading of all three books (Volume 2 in 1979, Volume 3 in 1980) by her friend, Elizabeth Chandler. In 1977 Thistle was 75 and Elizabeth was 68. This was a commendable achievement for two women in their later years, although Elizabeth claims that people in their sixties are at their peak. Thistle would stay overnight at Elizabeth's home in Carlingford, they would rise early, and with each one working on a separate copy, one would read aloud while the other checked, then change over. On each occasion the proof-reading was completed in about two days.

In September 1977 Thistle went to Queensland with Susan Heins, her illustrator for all three volumes in her latest series. The purpose of the trip was to collect specimens for drawings and data for material in the text. Thistle told her disastrous news to Kay in a letter. "Had a wonderful trip as far as Oakey (Brookvale Park between Toowoomba and Dalby) and rang the Gordons from there to say we would be out next day. Left B. Park after lunch and were travelling along at a leisurely pace (Susan driving) when a front tyre blew out and we rolled off the road. Over went the Galant, full of precious plants. Neither was hurt much. Susan had a fractured neck vertebra, two lower arm bones fractured (all apparently slight) and a knock on the head: I had a fractured collarbone, not discovered till the next day, and a few minor things. The car was terribly battered but the bloke at the towing station potted the plants for me (it was a Sunday) and a friend of Dorothy Gordon's was (and still is) minding them in Dalby. Susan had to stay in hospital for 3 weeks and Dorothy came out and collected me and I spent the time collecting seeds. The car is still not ready (it happened September 4) and I have to go back and get it — maybe the Swane's (from the nursery) will take me up and have a look at Gordon's but don't know when. I took a while to recover but am O.K. now. We were indeed terribly lucky."[34]

It was actually John Teulon from Swane's nursery who went back to Queensland with Thistle. They flew to Brisbane and then to Dalby, where John questioned the mechanic who had been working on the car, particularly to check on the steering. "There was a bend in the road and they didn't take it, that's all," said the mechanic. John drove the car down through Goondiwindi and back to Sydney.[35]

Susan Heins was an Austrian by birth and had taught art in a Technical College in Vienna. She had been friends with Thistle since meeting her in about 1945 and shared her interest in the bush and its plants, being a bushwalker and camper since her youth. She was a competent artist, and besides illustrating Thistle's later books, supplied illustrations on native flora for many issues of the Stead Foundation publication, *Wildlife Research News*. A reviewer of *Gardening with Australian Plants, Volume 3, Trees* wrote, "There are some pleasant, though not particularly distinguished coloured photographs. The line drawings by Mrs Susan Heins add more to the text than the black-and-white photographs which seldom reproduce well.[36]

A tribute to her friendship with Thistle was written by Susan Heins for the SGAP Newsletter 4.12.90 and reproduced in *Wildlife Research News* in March 1991 after Susan's death on January 18, 1991, in her 92nd year. Susan wrote in part, "I learned to know Thistle very well as we travelled together through parts of New South Wales, Victoria and Queensland and, of course the Snowy. I had the unique opportunity to see how everywhere there were people who knew her, greeted her with joy, invited her to stay with them and so we stayed, often in beautiful big homes but also in much humbler bush dwellings all over the country. In the city there were the people from the universities and botanical gardens, conservationists; in the smaller places there were local organizations like horticultural and garden clubs, Country Women's Associations, nurserymen and, of course, teachers and retired teachers who were once her pupils at the Teachers' College, all of whom she inspired with an appreciation of the natural environment . . . She was extremely energetic and very thoughtful, but sometimes she could be very abrupt to people she did not think worthwhile to spend time on."[37]

In the three-volume series *Gardening With Australian Plants* Thistle was once again drawing on the expertise of many specialists. She acknowledged the help of professional botanists, practical gardeners, librarians, nurserymen, horticulturalists, foresters, conservationists and classical scholars. Much of Thistle's knowledge was derivative, but her skill existed in being able to draw all the information together and present it in a manner which could convey the meaning to the newest beginner in the art of gardening. Thistle appreciated the willing help she received from others, but was it always "willing"? On July 31, 1978 she wrote to Dr Mary Tindale at the Botanic Gardens in Sydney to ask about *Acacia* distribution and discrepancies in publications. Dr Tindale responded to her request on August 8, 1978, but added, "I would be grateful if you would not send any more queries for some time, as I have an accumulation of requests from various people and scheduled proofs are expected soon.[38]

There was a third trip to Tasmania in February 1979, Thistle being accompanied this time by Helen Brown. Thistle planned to visit Cradle Mountain — Lake St Clair National Park, Ben Lomond, Hartz Mountain National Park, Mt Field and of course the friends she had made on previous visits. She invited Kay Geeves and Melva Truchanas to join them in her car for some of the excursions, saying to Kay, "If you can book us into all the interesting places I will certainly be grateful. Can I beg off any formal talks with groups this time? I'd like to see and chat with individuals but not groups, O.K.?"[39]

Thistle and Helen left Melbourne on February 9, 1979 and visited friends in Launceston before driving to Queenstown. "I have a memory of lots and lots of *Prionotes* on the Queenstown-Derwent Bridge Road which might be in flower with luck and it might be salutary for Helen to see Queenstown."[40] Then it was across to Cradle Mountain and Lake St Clair before going south to Hobart and Port Huon. Later they travelled up the east coast north to Coles Bay and across to Ben Lomond. A Tasmanian friend, Doris Wythes, had agreed to go to Ben Lomond with them; they met at a ski lodge in the park, Dorrie accompanied by her younger son Richard who came to help with the driving — it was a difficult road. They arranged to meet in their respective cars at the foot of the Ben Lomond road, where Richard tried to persuade Thistle to transfer to Dorrie's car, but Thistle was adamant — she would drive. So they set off, but a short distance along the road Richard said, "I won't go on until she gets in our car, she's not fit to drive up this road". Thistle finally agreed, and at the top she said, "I see what you mean". They didn't do much botanising that day because Thistle wasn't well, "She couldn't get to the top of the hills like she used to".[41]

The four days up the north of Tasmania were hectic and they arrived in Devonport exhausted. A busy day was spent with the publishers, Thomas Nelson in Melbourne, before driving back to Sydney. "Have been exhausted ever since. Had an even more hectic Board Meeting on Friday but am slowly recovering. It was a great experience to see you all again and discover that the same feelings of affection can remain unchanged with a long separation."[42]

By March all the work on the series of books for Nelson had been completed. *Trees*, the third volume, was scheduled to arrive in the week of March 17, 1980, and expected to be published in May. "It is a peculiar feeling to be a bit freer and I find it difficult to settle down, but I suppose I shall get used to it or find something else to do. There is plenty to do at Wirrimbirra, anyway."[43]

The following summer was dry and conditions at Wirrimbirra very depressing for Thistle. She wrote to Kay in August 1980 after a long silence. "Since I wrote to you (I had no idea it was so long ago) things have been up and down. First — last summer was hideous and poor Shirley our only ranger spent most of her time watering with drastic results elsewhere. We eventually got another ranger two and a half months ago. [Mark Nelson]. Just before he came we had a freak hailstorm — quite local, extending only a few km around us and only lasted about half an hour, leaving behind a very picturesque scene and enormous devastation. Most of the sarlon on the shadehouse went and some of the glass. The plants suffered very badly, both potted and in the ground. The trees were stripped and everything looks very sad. Slowly recovering but we have nothing to sell. We shall recover... Winter was very cold and following the long drought — and it is still dry — the spring will be poor for us."[44]

On the home front they were battling along too. Kel Stead, who lived in Boongarre with his wife, Ivy, suffered quite a severe stroke shortly after his retirement from the motor trade. Before his retirement there had been several minor strokes, none of them sufficient to keep him from his work. Towards the end of 1980 he also suffered several heart attacks; he also had cataracts on both eyes and could hardly see, but the doctors would not operate until his heart condition improved. Thistle remarked on how sad it was for him because, in addition to his loss of sight, his hands were so badly arthritic that he had difficulty using them for the handyman jobs that interested him. "Ivy of course is taking the greatest care of him (he is not bedridden and goes regularly for a walk to keep as fit as possible). Ivy uses her spare time painting the inside of my house while I do the nasty tiddly cupboards and odds and ends."[45]

Kay Geeves sent Thistle a copy of a newly published book in 1980 — *A Time to Care: Tasmania's Endangered Wilderness*, text by Norm Sanders and photography by Chris Bell. Thistle's response to this superb publication is a measure of her continuing concern over wilderness, and her readiness to still take action in such matters, even as she approached her eightieth birthday. She wrote to Kay, "The book has given me infinite satisfaction (coupled with despair as I realize that these idealists are in for many severe disappointments in their lives, but how stimulating to know they exist). It is so beautiful, both in text and illustrations. It has stimulated me to do what I have been putting off — writing to Ellicott (New South Wales)[46] and the Premier of Tasmania. Chris Bell must have landed himself deeply in financial debt. There was a splendid and sympathetic review in *the S.M. Herald* 27.12.80 I sent it to the authors. We should send copies to all Parliamentarians — but would they be moved?"[47]

The review of "A Time to Care" by Helen Frizell referred to "a more subtle campaign to preserve the island's wilderness areas. It's a campaign that's being waged with the help of photographs — the pictures of a former New South Welshman, Chris Bell. He has borrowed many thousands of dollars to produce his unique photographic odyssey through Tasmania's wild places." The reviewer considered that the photographs are even more eloquent than the text in pleading for Tasmania's south-west and central areas, for preservation of the alpine and temperate wilderness. In the Foreword to this book, Dr Bob Brown, as Director of the Tasmanian Wilderness Society, wrote:

> "Norm Sanders has written about the dilemma of a community which has failed to return the respect this Earth deserves. He points to the inadequacy of modern politics in its role as custodian of the eternal values of wilderness. This is the first book to look at that role and the responsibilities of Tasmanian politicians for the recent destruction of much of the island's natural glory. The gorge walls and forest giants have fallen to make way for their dams and loading bays, but so will their reputations fall in the judgement of history, in the eyes of our children for ever."

On June 14, 1980 Mrs Thistle Yolette Stead was appointed a Member in the General Division of the Order of Australia, the citation reading, *For Service to Wildlife Conservation*. Thistle was proud to be able to write A.M. after her name, even though she had earlier refused the MBE. The Wirrimbirra newsletter, *Wildlife Research News* September 1980, wrote a short piece "Our Congratulations Thistle" and added, "Those of you who know Thistle know also of her interest in conservation and education and of her untiring efforts in this field of work as well as her depth of knowledge in Australian flora and fauna". This award was fitting recognition of an active life spent in the support of a cause that had involved her body and soul. It was the culmination of an energetic and enthusiastic pursuit of her ideals as she entered a decade of relative inactivity and increasing debility. But first another visit to Wirrimbirra.

Chapter 9 Notes

1. *Wildlife Research News,* March 1971, p. 5.
2. *Australian Wild Life,* June 1972, Appendix.

3. ibid.
4. Hope, R. M. *Report of the National Estate*, 1974.
5. de Berg Tape 574, Side 2, p. 2.
6. ibid.
7. M. Truchanas, pers. comm. 19.1.97.
8. Letter from T. Stead to K. Geeves, 21.2.71.
9. ibid. 15.6.72.
10. E. Chandler, pers. comm. 3.1.97.
11. ibid.
12. J. Teulon, pers. comm. 3.6.97.
13. ibid.
14. *The Natural Garden* by Anne Latreille, p. 82.
15. ibid. p. 236 and p. 248.
16. E. Chandler, pers. comm. 3.1.97.
17. ibid.
18. Letter from T. Stead to K. Geeves, 21.2.71.
19. E. Chandler, pers. comm. 25.5.97.
20. Letter from T. Stead to K. Geeves, 6.11.73.
21. E. Chandler, pers. comm. 3.1.97.
22. ibid.
23. John Stead, son of Ivy and Kel.
24. Letter from T. Stead to K. Geeves, 6.11.73.
25. M. Truchanas, pers. comm. 19.1.97.
26. H. Rowley, pp. 506, 511, 513.
27. E. Chandler, pers. comm. 3.1.97.
28. Letter from T. Stead to K. Geeves, 14.10.74.
29. E. Chandler, pers. comm. 9.3.97.
30. Letter from T. Stead to K. Geeves, 14.10.74.
31. R. Mason, pers. comm. 23.11.96.

32. V. Serventy, pers. comm. 19.10.96.
33. E. Elenius, Introduction to *Bringing Back the Bush.*
34. Letter from T. Stead to K. Geeves, 6.11.77.
35. J. Teulon, pers. comm. 27.5.97.
36. *The Age*, 27.9.80.
37. *Wildlife Research News*, March 1991, p. 11.
38. Stead Papers, ML MSS 5715, 2(25).
39. Letter from T. Stead to K. Geeves, 1.11.78.
40. ibid. 11.12.78.
41. D. Wythes, pers. comm. 11.1.97.
42. Letter from T. stead to K. Geeves, 12.3.97.
43. ibid. 12.3.80.
44. ibid. 14.8.80.
45. ibid. 3.8.81.
46. Robert Ellicott, MHR (Liberal) was Thistle's local member in Wentworth, 1974–1981.
47. Letter from T. Stead to K. Geeves 31.12.80.

CHAPTER 10

Wirrimbirra Revisited — 1976–1990

"Drought is still with us and we are still having internal problems but I am hoping (as usual) that things will get better."

T. Y. Stead, December 1980

IN 1976 Wirrimbirra was looking forward, but as responsibilities and commitments were growing, hopes were also fading that the present economic climate would allow any government assistance to be forthcoming to meet the Sanctuary's needs. In the few years preceding 1976 the Councillors had been busy seeking assistance from various government bodies, both Australian and State. Long hours were spent in discussions of the things that merit the greatest attention (like finance), in compiling applications, in interviews and discussions. Some results were forthcoming: there was a grant from the National Estate to build two additional accommodation cabins; assistance with labour came from the Regional Employment Development Scheme (RED) when four workers were employed full-time for six months in 1974, good progress being made with the ground works during that time; the New South Wales Department of Education helped in the maintenance of study trails and in the continued services of teachers for the Field Studies Centre.

For the past few years the Foundation had been rejoicing in the knowledge that the background work carried out by the past generation of conservationists was at last receiving some recognition and that there was developing a strong public feeling and an awakening of public conscience in this field. "This has, for the moment been superseded by a feeling of apprehension for the future," was the fear expressed in *Wildlife Research News* in June 1976. Close to home, the Board at Wirrimbirra was concerned about sufficient finance to maintain its buildings, and pay for high labour costs. An ambitious developmental programme for the next few years ahead had to be curtailed, and it was necessary to consolidate what had already been commenced rather than undertake any additional commitments. The call went out for more voluntary assistance to help the Sanctuary through a difficult period.[1] A few years later when the freeway from Campbelltown to Mittagong was opened in 1981, traffic no longer passed regularly along the Hume Highway, and for Wirrimbirra this meant that the "world no longer went past the door". There was a considerable decrease in nursery sales and this meant reduced income.

Wirrimbirra needed a steady income to finance the day-to-day running and increasing maintenance. Thistle took what could be regarded as a drastic step to meet this need: in the early seventies she sold the block of land adjoining her house at Watson's Bay. The property left to Thistle by David Stead consisted of two blocks of land, each with their own titles, and the house which was built over both blocks. One of the blocks of land was sold, with payments being made over a period of time to enable Thistle to make regular donations to Wirrimbirra. This sale of land did not result in an immediate injection of funds for Wirrimbirra but rather enabled Thistle to increase the rate and size of her periodic donations over a number of years.

As Wirrimbirra had progressed, it was essential that the day-to-day management became more commercial in its direction. This didn't seem to suit Thistle in some respects because she did not have much commercial tact or business sense and found the consultation/negotiation unnecessarily tedious. There must have been mixed feelings for Thistle as she saw her dream realized in terms of development — two permanent staff, teachers supplied by the Department of Education, schoolchildren on site almost every day, groups visiting at weekends — but it was a long way from those early days at Wirrimbirra when a group of a dozen or so people would be there on a weekend with sleeves rolled up and happy to put up with minor hardships because of their common purpose to make the place operational.[2]

Changes were taking place in the personnel at the Sanctuary. Mel Fackender had been farewelled at the end of 1975. Bruce Knight, who had been appointed as Ranger in March 1975, was replaced in June 1976 by a young Victorian, Roger Thomas, and his assistant was Bill Mackay. Bill was to be replaced in 1978 by Shirley Rooke, who remained at the Sanctuary until 1990. Roger left in February 1979, returning to Victoria to take up a position as Ranger at the Clarkesdale Bird Paddock, Linton, near Ballarat. It had been a happy partnership — Roger the Ranger, and Shirl the Girl. The June 1978 issue of *Wildlife Research News* reported: "Shirley is a happy person, pleasant and competent, bringing goodwill, sound common sense and considerable ability to her job and we hope to continue our association for a long time". When Roger left, Shirley wrote for the *News,* "Roger's departure is being felt very keenly by me, and many customers to the Nursery, who have come to appreciate his pleasant personality and his extensive knowledge of both plants and birds, have expressed regret at his departure. However, the Bird Observers' Club have acquired an excellent Ranger, and we wish Roger well in his new position."[3]

On Roger's departure, Shirley became Ranger, but had to work on her own for a while as finances were low; the position became known as Manager. In the September 1979 issue of *Wildlife Research News* it was reported that a member of the Stead Foundation who wished to remain anonymous had offered to subsidize the salary of a second ranger for Wirrimbirra to the amount of $7,500 per year for two years. Therefore Mark Nelson came as Assistant Manager in 1979, and in 1981 Peter Cuneo, finishing off his BA degree at Macquarie University, took up part-time work at Wirrimbirra, mainly helping Mark in the nursery. In 1982 Peter was employed full-time, his salary being provided by donation from Thistle's personal finances. Mark left in 1983, and Peter became Assistant Manager.

The situation with management in the mid 1980s was rather sensitive. Shirley, very satisfied with her job, just couldn't seem to please Thistle, especially after Peter appeared on the scene. His intelligence and personality, and his keen interest in the nursery, appealed to Thistle, and she decided he would make a better manager than Shirley. In fact, Thistle had an important influence on the direction Peter took with his career. At first he didn't know how to take her — she was brash, and a bit eccentric, but very inspiring. His memory of her today is of an energetic figure, clipboard in her hand, cigarette in her mouth, and an old army-type soft hat on her head.[4] When Peter was working at Wirrimbirra, Thistle would arrive, driving herself, her car loaded up with pots of plants she'd struck up at home in Watson's Bay. Peter would unload, the writing on the labels often a challenge to read, and she'd stay one or two nights. He would then load the car again, this time with bags of potting mix, and empty pots. Her glasshouse at Watson's Bay was a haphazard affair, Peter said, and he was frequently an honoured visitor there, to be able to make a judgement. During her active years at Wirrimbirra the vehicles she drove were station wagons which always rusted out in the back because the plants on board were watered by Thistle *in situ* to stop the heat in the vehicle from affecting them.

Thistle had frequent disagreements with Shirley. Thistle liked positive, enthusiastic people who had opinions and commitment; she liked people who argued with her and stood up to her, and Shirley appeared to be too conservative. Thistle didn't respect Shirley even though the latter was a very good field naturalist and did a bit of nursery work too.[5] But Shirley had respect for Thistle, for her knowledge, and what she had achieved. In answer to a letter from the author in 1996, Shirley Rooke wrote, "Yes, I did respect Thistle. I respected her intelligence, her tremendous botanical knowledge. I respected her enthusiasm, her photographic skills, her writing ability. I respected her determination, unless it was carried too far. I respected her energy. I respected her kindness and compassion. She even let me practise learning to drive in her car."[6]

Without referring to the Board, Thistle instructed Shirley one day to plant waratahs along the Sanctuary boundary on the Hume Highway. A worried Shirley privately asked Ivor Wyatt what she should do. As both of them realized the plants would not last five minutes in such a public position. Ivor's reply was, "Do nothing". Fortunately that was the end of the matter.[7]

The Board was very satisfied with Shirley's management of Wirrimbirra, and wouldn't support Thistle's moves to dismiss her. Therefore, Thistle tried to make life uncomfortable for Shirley, in the hope she would leave. She would find fault with things Shirley did, or did not do. In the 1980s Thistle used to be very frustrated with both the management of Wirrimbirra, and with the Board, who would often meet in Sydney and function rather like an absentee landlord, quite satisfied with the status quo.[8] Keith Armstrong, Teacher-in-Charge at the Field Studies Centre for close on six years, said it was always interesting to observe from the sideline Thistle's dealings with the rangers and nurserymen. On arrival, each one would have a honeymoon as the golden haired boy/girl, followed by a period when it appeared that nothing was right. Some lasted longer than others, Keith said, but they were all wonderful to him and he valued their friendship and assistance. Having groups stay overnight meant some problems for the rangers as there was very little recreation area, mainly because Thistle wouldn't give up any precious areas.[9]

However, although Thistle would have liked Peter Cuneo to become Manager of Wirrimbirra, she realized such a position, with low wages and no further promotion, was not in his best interests, so when he applied for a position at the Botanic Garden at Mount Annan, she was quite supportive. Peter left Wirrimbirra in late 1986, and Thistle gave him a book. It was *Handbook of the flora of New South Wales, A Description*, by Charles Moore, assisted by Ernst Betche, 1893. In the front of the book she wrote, "A sentimental gift for Peter. My favourite plaything from 1918 when a friend of my father's 'borrowed' it from the Department of Lands. I hope you find something equally impressive during your lifetime. Thistle."[10]

On the education front, following Keith Armstrong's move to Orange at the end of 1978, Janusz Haschek became Teacher-in-Charge of the Field Studies Centre, with Mark Robertson as his assistant.

Prior to the Annual General Meeting in October 1979, the President, Allen Strom, advised that he would not be seeking re-election to the Board. Allen found it difficult to work with Thistle and had experienced similar difficulties when Secretary of the Wild Life Preservation Society. When Allen was asked why he agreed to become involved with Wirrimbirra, he replied, "I thought it was a worthwhile cause, and I thought the Field Studies Centre had great potential".[11] At the 1979 AGM, the following members were elected to the Board: Kathleen Groat, Frank Haddon, Colin Miller, John Rose, Thistle Stead, Chris Treneman, John Teulon, David Tribe, and Ivor Wyatt. At a subsequent Extraordinary General Meeting, Mr N. H. (Bill) Hicks was appointed to the Board and also elected President, with Ivor Wyatt as Vice-President. Kathleen Groat was the Honorary Secretary and Gilbert Stead the Honorary Treasurer.

In 1980, Patricia Meeks was the Editor of *Wildlife Research News* but in September 1981 there was a call for a new editor and there was an offer from Connie Burns and Doreen Clark to take the position as co-editors. Kathleen Groat retired from the position of Honorary Secretary in 1984 and Allen Strom wrote an appreciation of her dedication, efficiency, interests in nature conservation and contribution to the effort which made Wirrimbirra work. The teacher-in-charge at the Field Studies Centre was now Gordon Kovacevic. Office bearers on the Board for 1986–87 were Ivor Wyatt as President, Colin Miller and Thistle Stead as Vice-Presidents, Mary Cawte as Honorary Secretary and Gilbert Stead as Honorary Treasurer.

From early 1987 the new Assistant Manager was Howard Duff. Howard had studied at the University of New South Wales and Ryde School of Horticulture, completing his studies in 1981. He became a casual gardener at Wirrimbirra in 1983 and worked there for eighteen months before spending two years in Hobart doing landscape work. Thistle wrote to Kay Geeves in April 1987, "Howard Duff came up to see me a day or two after his arrival in Australia (Wirrimbirra). He is full of sensible ideas. He and I are going to rehabilitate Wirrimbirra Gardens which are in a sorry way. Alas, W. is now just another native plant nursery. I am hoping Howard will help bring it back to its original purpose. I can't do much." She was then 84 years old. She explained in a later letter that with help from one or two Board members and with Howard's help, they could resuscitate the conservation side in addition to the school.[12]

Unfortunately Howard only stayed twelve months, leaving to live in Victoria. His replacement was Philip Harrison who had previously been involved with the "Trees for Farms" organization.

The Foundation received a bequest of $1,000 under the will of Jean Stevenson who died in 1987. She and her friend of many years, Nell Reeve (who died in 1988), were both Life Members of the Foundation, and were active in the establishment of Wirrimbirra Sanctuary from the time that Carmen Coleman paid the deposit on the proposed site at Bargo.

In February 1988 President Ivor Wyatt prevailed upon the National Trust to purchase a strip of land behind the cabins and schoolrooms, stretching from the main road to the railway line. The purchase price was $7,500 but with fencing and fees (legal and surveying) the total cost was almost $10,000. The objective was to have land in which to create a fire break — selective clearing of timber could be carried out — but there was the possibility of being able to create a barrier if the adjacent land at any time were to be occupied by unsympathetic development.[13]

The Women's Committee of the National Trust visited Wirrimbirra in October 1988 and spent most of the day inspecting the facilities and hearing about the activities of the Foundation. No doubt impressed by what they had learned, the ladies agreed to make a donation of $1,000.

For the year 1988–89 the Board was unchanged, with Ivor Wyatt still in the position of President. Julia Robertson had become the Minutes Secretary, the Honorary Solicitor was D. Webber and the Honorary Auditor, R. Hazell. The March 1989 issue of *Wildlife Research News* reported on an address to the 1988 AGM by Mr Bill Hicks, first President of the Foundation. He said, "If the Stead Foundation achieves nothing else its special contribution in the field of environmental education will be an outstanding memorial to the two people who brought the Foundation into being, Carmen Coleman and Thistle Stead, and to Allen Strom who guided the Environmental Field Studies Centre into being". Bill Hicks also mentioned the contribution of Mr Ivor Wyatt who, together with Thistle Stead, were two members of the Board who had given uninterrupted service to the Foundation since its inception.

In early 1989, Mr Stephen Benoit, who had been teaching at the Field Studies Centre since the departure of Mark Robertson in 1988, became teacher-in-charge, as Gordon Kovacevic had been transferred to Dorrigo. Philip Harrison returned to Victoria, and Paul Davis was assistant to Mr Benoit.

A small but dedicated group of workers, the Friends of Wirrimbirra, had its first meeting at Wirrimbirra on February 12, 1986. By early 1988 there were only three members. During 1986 Mr and Mrs David Hunt of Thirlmere commenced canvassing their neighbourhood to ascertain the local interest in the formation of a Children's Friends of Wirrimbirra group. Encouraged by the response, a group was subsequently formed — the Wirrimbirra Wanderers, a junior group of the Friends of Wirrimbirra. Their activities in 1987 were:

- regular evening club meetings every second Friday of the month at Wirrimbirra.
- weekend camps or day trips every three months.
- periodic Sunday activities at the Sanctuary.

Under the heading "Quiet Contributions" the March 1988 issue of *News* recorded the appreciation of the Board to Mrs Ivy Stead, who for over twenty years had packed and dispatched the quarterly *Wildlife Research News* and other Foundation publications, maintaining a liaison with membership records. Long-term members, the *News* added, will recall Ivy's hospitality "behind the teapot" at earlier AGM's and on other occasions. Appreciation was also expressed to Miss Kate Stead who for about fifteen years had addressed envelopes for the quarterly *News* and made alterations, as necessary, to membership records. Fortunately, both ladies were continuing their voluntary work for the Foundation.

In the President's Report for 1987–88, Ivor Wyatt referred to another "Quiet Contributor" — his wife Mabel, "who copes with a tremendous amount of typing of reports, agendas, correspondence, minutes, etc.". Ivor also paid a tribute to Colin Miller, "Colin, a Vice-President and Chairman of the Management Committee, gives a tremendous amount of time to the Sanctuary and its affairs, is here very frequently, and assists in so many ways. He and I work as a team and I cannot speak too highly of his valuable contribution to the Foundation."

Open Day at Wirrimbirra in 1988 was of special significance as it marked the official Bicentennial project concerned with the propagation of the "Wirrimbirra White" waratah. Thistle told of her excitement in hearing of a white waratah growing on private property in the Colo Vale area near Mittagong. She visited sometime in the early 1970s and spoke to the rather obstinate owner who refused to allow any material to be taken from the plant; a vicious dog was tied up near the base of the plant. As the plant was old and spindly Thistle advised strongly against transplanting which was being considered due to its proximity to a new road. She spoke in favour of hard pruning and rejuvenation, but the transplanting took place and the plant subsequently died.

Not long after this episode, a Water Board employee had seen and picked a flower of the white waratah from a plant he had found whilst working in the southern catchment area near Kangaloon. This beautiful flower was then displayed that evening on the bar of the Thirlmere pub. News of this rare flower spread to Wirrimbirra, not far from Thirlmere, and Thistle immediately drove over to see it. In her persistent manner she began quizzing the somewhat reluctant Water Board employee who eventually told her the approximate location of the plant. Arrangements were soon made with the Water Board for Thistle to visit the area, collect cuttings and propagate this rare form.[14] Early attempts to strike the cuttings were unsuccessful, but in time many plants were obtained. John Teulon was asked by Thistle to try his skill with some cuttings. He recalls that about 1973 he went to Bargo and was taken by a Water Board employee to the site of the rare plant. He took about ten cuttings, which he treated with hormones, and about 80% took root and survived. He handed them over to Thistle who gave one back to John, sent others to the Botanic Gardens in Canberra, and retained the rest for Wirrimbirra.[15] At least as late as 1993 the original white-flowering plant was still in the Water Board catchment area, closed from public access, but cuttings were no longer being taken from it as there were then sufficient plants in cultivation.

In 1986 Thistle was successful in having the white waratah registered as a cultivar with the Australian Cultivar Registration Authority. It was registered as *Telopea speciosissima* "Wirrimbirra White." All the white flowering plants now in cultivation are believed to have originated with the plants originally cultivated by Thistle.

By 1988 many plants of the white waratah had been established through judicious distribution to skilled propagators, though there were still insufficient plants to make them available commercially. To date the Botanic Gardens in Canberra had been the most successful because the soil there was much more suitable for the growth of waratahs than that at Wirrimbirra. The President of the Foundation, Mr Ivor Wyatt, explained at this 1988 Open Day that with the assistance of the Bicentennial Authority, an attempt had been made to increase the possibility of greatly expanding the availability of specimens by commissioning a scientist to propagate the white waratah by tissue culture. The Foundation received a grant of $6,000 from the Australian Bicentennial Authority in 1985 and invited Dr R. de Fossard to conduct the research into tissue culture in his laboratories, then in Armidale, New South Wales and later in Tamborine, Queensland. $3,000 of the $6,000 was advanced for Stage 1 of the research in 1985. At the Open Day in 1988 the report was, no success to date, and Thistle was reluctantly advising that the technique be abandoned for an alternative method. However, a plaque to commemorate the Bicentennial project was unveiled by the local Member, Honorable John Fahey, M.P., in the presence of the local Councillors, Shire Clerk, visitors and members.

In the 1990s several attempts with tissue culture were successful, but it has never been a commercially viable enterprise. In tissue culture experiments to date, the specimen tissue develops roots, but in the hardening off period the plant rarely survives, i.e., in the transfer from the laboratory to a natural environment the plant does not survive, often succumbing to disease. While the white waratah is not sterile, obtaining seeds which breed true to the white is not easy, and propagation continues to be mainly by cuttings.[16]

Assistant Manager Philip Harrison left his position at the Sanctuary early in 1989 and was replaced by Matthew Fowler who also left after a short stay, at the end of the year.

On Sunday, October 29, 1989, in fine warm conditions, between 750 and 800 people visited the Sanctuary for its 1989 Open Day. The nursery was the focal point for many visitors. Board member Laurie Greenup gave an enlightening talk on photographing plants. This included hints on selection of cameras and film for amateur and professional photographers. The highlight of the day, however, was the inauguration of the sale of "Wirrimbirra white" waratah specimens as advertised in previous issues of the *News*. The President, Ivor Wyatt, acknowledged that the present situation in which plants could be made available for home gardens was due to the dedicated and determined efforts of Dr T. Y. Stead, over many years, to preserve the only known specimen of white waratah in existence and to ensure that propagation would provide the means for the continued existence of specimens of this rare plant for future generations.

The President, Ivor Wyatt, invited Mrs Noila Berglund, Director of Education, Metropolitan South-West Region, to inaugurate the release for sale of the "Wirrimbirra White". Mrs Berglund described the Sanctuary as a place which provided an opportunity for expanding experience of the natural environment for children who might otherwise be almost completely deprived in this respect. Such experience might lead on to permanent interest in adult life. Most appropriately she chose to allow the "Wirrimbirra White" to make "its debut", rather than "launching" or "opening". Mrs Berglund went on to announce a special grant of $3,800 from the Region to enable the Foundation to carry out very necessary painting and minor repairs to cabins 4 and 5. The President, aware that such a generous gesture deserved acknowledgement, immediately put his arm around the guest, announcing "You are a lovely lady", to the applause of those present. By March 1990 the painting and repairs had been completed.

In early 1990 Shirley Rooke resigned as Manager, effective from 27 January, in order to reside closer to members of her family and to occupy her nearby home in southern Queensland. Her position as Manager was taken by Philip Harrison who had been her assistant from April 1988 to January 1989. The Board entertained Shirley at a special luncheon and made a presentation on behalf of the Foundation.

A generous bequest of $81,551 from the Oxley Estate was made to the Foundation in 1990. The Board had been aware for some years that the Foundation would ultimately benefit from the will of the late Jean Molesworth Oxley, but had no idea of the amount nor when payment could be expected. A complicated legal situation concerning various Oxley family estates prolonged finalization of this Estate. The late Miss Oxley joined the Foundation in 1964 some two years after its establishment, becoming a Life Member in 1969. She gave financial support prior to her death in August 1972.

On July 5, 1990 Dr Thistle Stead passed away as a result of a heart attack. *Wildlife Research News*, September 1990, issued tributes from four long-time friends. One paragraph in the tribute written by Bill Hicks is a succinct summary of Thistle's relationship to Wirrimbirra.

> "The Stead Foundation and Wirrimbirra Sanctuary will not only perpetuate the memory of David Stead but will be a continuing memorial to Thistle Stead because it exists today not only as a result of her original initiative but through her continuing drive and generous financial support over many years."

True to her commitment as a scientist, Thistle offered her remains to the University of New South Wales Medical School for scientific research — there was no funeral, there is no tombstone. And true to her twofold commitment, David Stead and the natural environment, she made provision in her Will for the future of David's memorial, Wirrimbirra. Thistle's Will, drawn up in October 1986, was explicit: "It is my express wish that all benefits received by the National Trust of Australia (New South Wales) under this my will, be applied by it towards the development of its property presently known as Wirrimbirra Sanctuary, Hume Highway, Bargo". Thistle's single-minded concern over Wirrimbirra led to her disregard of the family as the bequest of almost $1.2 million was left to The National Trust. However, certain members of the family, as well as many friends and followers, believe the Trust did not fulfil its obligation to Wirrimbirra, even taking action in 1995 in the Equity Division of the Supreme Court to determine whether a true construction had been placed on the will and its bequest to the Sanctuary. But later the case was withdrawn as the Stead Foundation came to an agreement with the National Trust. As the nineties progressed there was no solution to the stand-off between the opposing factions, but for supporters of Thistle and her dream, the hope is still there that a solution can be found, that Wirrimbirra will indeed survive in perpetuity.

By 1998 this hope was moving into reality; plans were being discussed for a new Visitor Centre, upgrading of the cabins, and further consideration of involvement in the area of education and public programmes. On November 23, 1997 the first part of the new vision for Wirrimbirra was realized with the opening of the Fauna Protection Area, the enclosure for local indigenous wildlife, a project to which the National Trust had contributed as part of its agreement with the Foundation. The 1998 Board of the Stead Foundation is committed to Thistle's dream and President Al Mucci expresses it clearly — "As President, I will continue Thistle's lifelong ideals for Wirrimbirra and with the Board will continue to achieve these ideals for the benefit of Wirrimbirra".

Chapter 10 Notes

1. Information from *Wildlife Research News*, June 1976, p. 5.
2. J. Stead, pers. comm. 14.7.97.
3. *Wildlife Research News*, June 1979, p. 8.
4. P. Cuneo, pers. comm. 29.9.96.
5. ibid.
6. Correspondence from S. Rooke, 20.10.96.
7. I. Wyatt, pers. comm. 8.7.97.
8. P. Cuneo, pers. comm. 29.9.96 and 30.4.97.
9. K. Armstrong, pers. comm. by letter, 14.5.97.
10. P. Cuneo, pers. comm. 29.9.96.
11. A. Strom, pers. comm. 2.1.97.
12. Letter by T. Stead to K. Geeves, 10.6.87.
13. I. Wyatt, pers. comm. 8.7.97.
14. Notes supplied by P. Cuneo.
15. J. Teulon, pers. comm. 3.6.97.
16. P. Nixon, pers. comm. 10.6.97.

CHAPTER 11

Declining Years — 1982–1990

"Time grows precious now."

T. Stead, June 10, 1987

IT was 1982 and Thistle was in her eightieth year, a committee member for The Year of the Tree. "It is a large committee, very enthusiastic and with some good people but the direction of it is not good and Government has not come forward with any money yet. So we jog along, hold laboriously long-winded meetings, and wait. It becomes very depressing."[1]

Earlier in the year Thistle suffered two setbacks. The first was a car accident when a truck ran into the back of her station wagon and pushed it on to another car in front. Her car was a write-off, and even though the insurance company would have repaired it, they did not recommend it. She bought a Honda Accord hatchback which was difficult to load but a delight to drive, and was quite pleased about the outcome of the accident as she was not at fault and was repaid the whole of the insurance on the wagon. A short time later she had another accident — ignored a step at Swane's Nursery, fell, and fractured several ribs. "Very painful. I used water bags and heat lamps and ointment and painkiller and I don't know which were effective as I used them all at once. It is now practically O.K. I discovered too that not being able to stand or move about much for more than two hours at a time I had to rest and this did me a great deal of good — quite rejuvenated me."[2]

In the early 1980s Thistle was still driving herself around, and apart from the long trip to Wirrimbirra, regularly undertook the relatively long trip from Watson's Bay to Wahroonga to visit her sister Joan. Thistle's younger sister Joan had multiple sclerosis and was in a wheelchair for twenty-five years before her death in 1991. The two sisters had a good relationship, although in earlier years when Joan and her family would occasionally lunch at Doyle's restaurant in Watson's Bay, Thistle would never join them — she would meet them after lunch. This was not a slight to her sister, but a protest against the Doyle family whom Thistle believed had betrayed their working class background by catering to the wealthy. In the early 1980s, when Joan was in Neringah private hospital at Wahroonga, Thistle would arrive with a supply of fruit, take Joan into the garden and sit with her for a while, as long as there was no chit-chat.[3]

Time was dragging for Kel Stead as he had another heart attack later in the year, and two operations on his cataracts. His arthritic hands were "a major disaster as though a splendid man with his hands and kindness itself he has no interests other than those associated with such manual work or associated things. This makes him very difficult to live with and poor Ivy is near distracted. She gets some relief in hard physical work and the house falling to pieces from old age has never looked better. She has painted the whole of the outside and you know what a size it is. My own troubles are slightly less. Wirrimbirra still presents some problems — worse in some ways but maybe just as well as I have thrown down the gauntlet."[4]

Christina Stead died on March 31, 1983, aged 80. The final years of her life spent in Australia were not happy ones. She had recurring heart attacks and drank too much alcohol, but the main problem was her loneliness, in particular the presence of a stimulating male companion. The domestic tension at Hurstville, where she lived with her brother Gilbert, was distressful. When she left for ten weeks as Writer-in-Residence at Monash University in Melbourne in 1977, she resolved never to return to Hurstville. At 75, without her husband Bill, and in poor health, she was a wanderer again. From March 1977 until her death in March 1983 — six years in all — she would be restlessly moving from place to place. In December 1978 a letter came from the

Official Secretary to the Governor-General offering to make her a member of the General Division of the Order of Australia. She declined.

No one in the family wanted to have Christina permanently; she was not an easy companion. At one point Christina asked Thistle if she could live at Boongarre. Kel was in very poor health at the time and Thistle (whom Christina had ignored all these years) had no illusions about David Stead's eldest child; she said no, and felt bad about it for years. Christina's last home was with a young doctor, Helena Berenson, who lived in a rambling Victorian house in Johnson Street, Annandale, with her husband Geoff Davis, also a doctor. Helena became Christina's nurse, doctor and friend. One night in March 1983 they found her wandering about delirious — it could have been a combination of fever, dehydration and alcohol consumption but Geoff Davis believed there was something else. They took her to Balmain Hospital where she died a few weeks later. Only seven people attended the cremation at the Northern Suburbs Crematorium, among them David and Doris Stead, and Gilbert Stead, the only family members. Her sisters were not there — she had said she did not want them to attend — and friends were dismayed by the bleakness of the occasion.[5]

In June Thistle was in plaster with a broken ankle bone, but she had been released by the spring when Peter Cuneo, Assistant Manager at Wirrimbirra, celebrated his wedding. "His wedding was delightful. The big *Pandorea pandorana* was scheduled to give a wonderful background but [there was] unseasonably hot weather shortly before that so instead it was a white carpet. But it was all very charming and a very happy, jolly crowd. The reception hall was decorated with wildflowers by Peter's grandmother and the cake also decorated with wildflowers. We had a good A.G.M. this year too — very happy and friendly — Peter and Kim came back halfway through their honeymoon to attend and then shot off again only to appear a few days later (still on their honeymoon) to have another look at their house. They have moved into the big cottage [at Wirrimbirra] and are loving the fun of getting some extra furniture."[6]

Sometime in 1983 Thistle prevailed upon John Teulon to travel with her to Dripstone to see the Althofers. John was much pressed at work and was reluctant to go away for the two or three days required. He could only spare one day, he told Thistle, and that would mean too much travel on one day, but she agreed, and arrived at John's home in Dural about 5 a.m. ready to travel out west. John did most of the driving, but going home she insisted on taking the driver's seat, swerving on more than one occasion to the wrong side of the road. Thistle said to John, "You'd think they'd have a marked centre line on these main roads." John said, "They do, but it's on my side". (He was the passenger!)

Just out of Bathurst, and it was dark, Thistle was tired; when a semi-trailer passed them as it approached, but on the left side of John in the passenger's seat, he said, "I'll drive, I feel sick". "Oh," said Thistle, "if you're car sick it will be better for you to drive". Home in Dural, John's wife Judy persuaded the weary eighty-one year old to stay the night. Next morning Thistle said to Judy, "I didn't realize John got so car sick".[7]

It was in July 1983 when Kay Geeves was in Sydney with her husband and her son that Thistle insisted on driving them to the airport. She was in animated conversation about the plants along the roadside when she drove through a red light and just missed being struck by a huge semi-trailer. Kay's husband said, "Never again, even if a taxi costs $80!" The next year when they again visited Sydney, Kay was on tenterhooks about Thistle and her driving because Kay was a lady who found it difficult to say, No, to Thistle. On visiting Thistle, that lady said to Kay, "I've got bad news, the traffic has been so bad, I've handed in my licence". Kay breathed a sigh of relief.[8]

Thistle's activities were much curtailed in the 1980s but one event which gave her pleasure was attending a ceremony at Balgowlah Heights Infants School in 1984 to "cut the ribbon" at the Thistle Y. Harris Native Garden in the school grounds. The project dated from the 1950s when Douglas Piper, the Principal of the Primary and Infants Schools, emphasized curriculum study of the flora and fauna in the natural bushland on the vacant land across the road, the site of the future Infants School. Each week, study was made of a different plant or bird indigenous to the area, thus engendering in the children a total appreciation not only in that area, but in Australian native flora and fauna. The reference books used were Cayley's *What Bird is That?* and Thistle Y. Harris's *Wild Flowers of Australia.*

So successful was this "live" approach to the environment that Douglas Piper decided to create a Native Garden in the area, call it the Thistle Y. Harris Native Garden, and invite Miss Harris, Members of Parliament for the area, Council dignitaries and representatives from the Department of Education to attend a ceremony of dedication. The garden flourished for many years, but owing to changes in the administration, was not given the same care. Also, a new building to house a separate Infants Department was erected on the stony, soil eroded section of the site. The Thistle Y. Harris Garden remained intact but in need of attention. In 1973 Mrs Vivienne Webb became Mistress and, appreciating the beauty of the area and intending to use it as a medium for Nature Study, she set about to restore the area with the help of staff and parents. On November 21, 1974 the Garden was re-dedicated at a splendid ceremony with speeches, song and dance. Doug Piper attended with Norm Drummond, who had been Minister for Education in the fifties. Among the parents were former pupils of Mr Piper who remembered their school days in the Native Garden. Thistle attended also and was delighted with it all. Later part of the area was converted into an outdoor learning area for drama, music or craft. Part of it became a sheltered lunch area. In 1984 the area was again improved with a $3,000 Commonwealth Employment Scheme Grant and Thistle returned yet a third time as guest of honour at a special ceremony. Mrs Vivienne Webb, now Trustee of the Garden, was responsible over the years for saving the area from being (1) a football field, (2) a swimming pool, (3) a turning circle for parents' cars, (4) an access road. In the late nineties the area is still there, but not in the same pristine condition as in the eighties, nor is it being used with the same interest.

However, a few parents appear to be interested in restoring the lunch area.[9]

A successful open day at Wirrimbirra in September 1984 was a great morale booster for Thistle. More than 500 people attended, coming and going all day, the Highway lined with cars after every available parking space in the Sanctuary was taken up. The crowd was contented, and no one grew impatient with the queues at the nursery, bookshop and barbecues. However, Thistle did complain about the organization; even the President had to go shopping twice for bread, no one had been given the responsibility of looking after supplies of tea and coffee, but somehow they materialized.[10]

"I am seriously thinking of submitting to an importunate publisher who is urging me to allow him to publish a one-volume 'Gardening with Australian Plants'," Thistle wrote to Kay Geeves in January 1985. "He is keen to call it 'The Thistle Harris Book of Australian Gardens', or something like that. I resist — it sounds pretentious to me, but I admit a good selling point (the use of the name). He (the publisher) has bought the rights (to publish). He is holding me to a contract (or will be if I sign) to complete the re-arrangement in time for publication before Christmas 1986. This sounds a long time, but I think it will take more time than a new volume. He is negotiating to purchase the film (of the manuscript) from Nelson and rephotograph it. I am not sure this will work — and the purchase price is very high — the alterations would be considerable and this bumps up the cost of production and of course the selling price. There have been so many new plants on the market and so many alterations to names and groupings. Margaret Barrett who edited my last three, and is free lance, will edit this and will, in addition do a lot of the hack work which I and not an editor would normally do — checking names, writing letters, etc. She has some botanical knowledge. I did play with the idea of a co-author but don't know how it would work and really don't have anyone in mind. There will be 300 additional line drawings. I hope the publisher will agree to use Susan [Heins] again. She and I agree that at 86 she is doing better work than she has ever done, using a hand lens."[11] By August 1985 the "importunate publisher," who had offered unbelievable rewards, had decided not to go ahead with the project, part of the problem most certainly being the high handed demands from Nelson. Thistle was quite relieved.

Kay Geeves had been urging Thistle to take a holiday with her in Tasmania, but she was reluctant to leave home while Kel was still unwell — he had suffered another heart attack. Furthermore, her own health was not good. "I am uncertain of the future. I had some severe internal trouble accompanied by an external very itchy rash. The doctor finally whipped me off to hospital (apparently he feared a carcinoma but there was none). I had some most unpleasant tests and the diagnosis was finally shingles, as secondary symptoms. This was several weeks ago and the secondary effects are still in evidence and with this peculiar disease anything can happen."[12] However, this indisposition was not serious enough to prevent her accepting an honour from the University of Wollongong in August of 1985.

Thistle received a letter from the Vice-Chancellor of the University of Wollongong to say that 1985 marked the University's tenth anniversary and a special graduation ceremony was being organized for Friday, August 9, to commemorate the occasion. "In connection with the celebrations the University recently considered your contribution to teaching and research in natural science, particularly in the botanical field and in wildlife preservation. The University is particularly conscious of your lifelong commitment to science and to conservation and to the outstanding contribution you have made to our national life in these spheres. In recognition of this contribution, the University would like to confer on you the honorary degree of Doctor of Science."[13]

This letter from an academic institution gave Thistle deep satisfaction, because she saw in it a gesture which would surely raise the status of natural environment conservation in the eyes of the public, and in the outlook of academics also.

The University sent a car to Sydney to take Thistle to Wollongong in the company of Sir Roden and Lady Cutler, the former to receive a Doctor of Letters in the same ceremony. They were taken back to Sydney the next day after a night in a high-class hotel, which had followed a Tenth Anniversary Dinner as part of the celebrations. Thistle wrote to Kay Geeves, "But best of all I found that life on the campus at Wollongong is very different from that of others I know. I was not acquainted with W. before but they are such a jolly, happy, fun-loving, friendly crowd that instead of being a bore it was a lively and satisfying experience".[14]

An excerpt from the citation at the University of Wollongong reads:

> "When she first began her work of educating Australians about their own environment half a century ago, there seemed to be sufficient natural bushland still remaining to allow for study and recreation for the future. In those days Thistle Stead was among the few in the academic and general community who recognized the necessity for the retention and preservation of the natural environment. Today there is a reversal of that situation; there is a scarcity of untouched bushland available for research or enjoyment, and even that is threatened by development; but there is much greater appreciation of its importance by a larger number of the community . . . the University is paying homage as well to the cause for which she worked with such distinction, the sustaining of our natural heritage."

In December 1985 Kel Stead was in hospital waiting for investigations into a possible by-pass, as his heart had rejected the pace maker which had been inserted. Ivy was visiting the hospital twice a day, in between conducting her Christmas festivities.[15] Thistle was not without her problems. She had painful arthritis, or whatever, in both legs and both arms, and was having therapy, "much better, but only at the expense of two hours massage, exercises and hot pads which doesn't leave much time for other things when one crawls so slowly.

Added to this I had great alimentary discomfort and a general feeling of 'all's not well' . . . I am still seeing a urologist (who came out here 29 years ago to take blood from David and said that when he left he went and bought a copy of *Wild Flowers of Australia*). That's fame for you. I think — and so does my G.P. though he doesn't tell me — that there is a very definite mental factor in this and I think also that it dates back to just over two years ago when a very pressing publisher had offered me the most unbelievable contract to produce a one-volume 'Gardening with Australian Plants'. . . I spent some three months working really hard on the new text (which wasn't easy) . . . Before negotiations were complete there was a big economic slump forecast and big rises in the price of paper and printing. The proposed publisher reneged. I was very angry about this but on further reflection I think both of us got out of a very nasty situation." This was a letter to Kay Geeves in March 1986.[16]

Walking down the stone steps to Watson's Bay beach with two friends, Thistle tripped one day two months later and needed three stitches in a cut above the right eye. Her legs were giving no trouble except when getting up, and sitting for longer than twenty minutes. She was unable to bend to weed or do useful things like potting; she was doing little more than watering. But Kel was home from hospital after a very successful heart operation, and was considerably more cheerful.[17]

But Thistle Stead was not bored. She spent a considerable amount of time in 1986 sorting her negatives, transparencies and prints, and getting her books into some semblance of order so that they would go to the right people. She was trying to persuade Woollahra Municipal Council to save enough money from their park and street plantings to at least water the many plants they put out each year (badly planted in her opinion). The attempt to combine her Garden trilogy in one volume had not been abandoned — she wanted to do it in her own way and in her own time without a publisher hurrying her along, but she nevertheless had her doubts about achieving her goal. She was also trying to persuade Woollahra Council to operate a citizen involvement scheme to improve the area. Another major task was collecting enough data to have the Watson's Bay Post Office declared an Historic Building to avoid its threatened closure on January 31, 1987. There was not much hope of success in this quarter because her research to date (March 1987) showed "no historic significance" and "no architectural qualities". However, she was still completing a nomination form for the Post Office to be listed as an Historic Building.[18] On February 5 Thistle typed a letter (with many typing errors) to send to Bob Hawke asking for his consideration on the proposed closure of the Post Office.[19] In early 1987 Thistle reported that she and her supporters had been successful in keeping the Post Office open, but post officials were appealing against the decision.

Thistle at 84 years of age had not given up writing letters to the newspapers. In response to an article in the *National Times on Sunday* for November 30, 1986, a commentary on four aspiring authors preparing a biography of Christina Stead, Thistle wrote to the Editor on December 5. In spite of Christina's unsociable behaviour towards Thistle after her return to Australia, Thistle still had a high opinion of Christina's abilities. In part her letter reads, "Christina Stead's biography will not be an easy task. From the turbulence of her childhood to the devoted and concerned husband who admired her work and helped her [so] frequently she emerges as a writer unique in her own field. Her extraordinary ability to present in words the characters she portrays, including herself (whom she does not spare) and other recognizable relatives, friends and acquaintances marks her as a biographee par excellence."[20]

In May 1987 Thistle enjoyed a day at Wirrimbirra with the Friends; it was work in the morning followed by a talk from "one of our old Wildlife boys" on selecting, planting, caring for trees. The talk was well received, with many questions and much discussion. Thistle was now feeling much happier about the situation at the Sanctuary. She also went out one morning in Sydney to enjoy breakfast at a Chinese restaurant, "Quite fascinating to watch the great crowd eating hugely, but an awful waste of time. Time grows precious now."[21]

Wirrimbirra, its management and its future, was always in the forefront of Thistle's thinking. When the Bicentennial Authority and BHP (Broken Hill Proprietary Company Ltd) organized a project they called Pursuit of Excellence to reward persons who excelled in various fields, Thistle decided to enter in the category of Environment. The successful applicant would have the opportunity to nominate an organization working in that area to receive a grant, and Thistle could see a benefit for Wirrimbirra to the tune of $40,000. Thistle asked her friends and associates to support her application in writing, and this they were pleased to do. This was in October 1987. In January 1988 she wrote to Kay Geeves to say she had missed out on an award. "I had quite forgotten my rather ferocious attacks on BHP (in the 30s I think) when their station a little south of Wirrimbirra was doing some reckless anti-conservationist project and I attacked it. You can't win that way."[22] The time she mentions as being in the thirties is somewhat doubtful — Thistle was now suffering from Alzheimers disease and her memory had sadly deteriorated. "I think I probably told you about my blackout," she wrote to Kay. "None since. Have to go warily which I hate . . . If the pills I am taking for heart do the trick I will be O.K. but have to take them for ever."[23]

A family member believes Thistle was in dementia as early as 1987, or before that. Certainly she was confused over some of her details when interviewed in November 1987 for the New South Wales Bicentennial Oral History Project. "When did you get married?" Thistle was asked by Catherine Johnson, the interviewer. "I think it was 1946 — no, much later than that — in the sixties will do." She married in 1951, and David had died in 1957.

John Stead has said that the family put some of her erratic behaviour down to the fact that she was frustrated at her lack of mobility, this dating from the time she had been unable to renew her driver's licence. Always a smoker, she caused several minor fires (bed and wastepaper basket) which, in a weatherboard house were a real cause for concern.[24]

In the last few years of her life at Watson's Bay she would go wandering about the neighbourhood, brought home often by perfect strangers. The local shops were asked not to sell goods to her as her requests were nowhere within reason. Kay Geeves had written suggesting she should move to a care situation where she'd be looked after and have her own room to meet her needs. She replied that an important factor kept her at 14 Pacific Street, and would continue to keep her there. "This place belongs to David and me and I could not leave him. We were one person, so how could I leave the place."[25]

In August 1989 Thistle was assessed by a Sydney hospital and part of the resultant report read:

> "There is clear evidence of dementia of a moderate degree. There are defects in perception (disorientation in time and place); in memory (0 out of 6 events for short term memory, significant defects in distant memory e.g., details of family circumstances); . . . limited botanical skills despite this being her expertise, . . . I assume this is Alzheimer's Disease and it has been developing for the last four or five years. Things became obviously a little difficult some twelve months ago and have come close to being desperate for this last month . . ."[26]

By May 1990 Ivy could no longer cope with two invalids and reluctantly agreed to family pressure to place Thistle in a nursing home at Summer Hill. David D. went to visit her and she told him, "Someone stole my car". She hadn't driven a car for years. David said, "Did you get it back?" and she answered, "Yes". But her friends Connie Burns and Doreen Clark visited her in the nursing home and she always knew who they were. Connie says that Thistle was a great believer in euthanasia and in her declining years was very frustrated because it wasn't available to her. Five weeks after entering the nursing home Thistle Harris Stead died from a heart attack on July 5, 1990, a few weeks short of her eighty-eighth birthday.

Thistle Harris Stead

I set out to find the facts about the life of Thistle Yolette Harris. I found her to be like many of us, a contradictory character: dedicated, passionate, and intolerant; energetic yet slapdash; a user of persons, yet a facilitator for their talents; a mediocre scholar, yet the supreme educator. She was forthright, courageous and charismatic, but turned aside from those whose opinions or beliefs were in opposition to hers. Her life was controlled by two passions — her love for the Australian environment, and her love for David George Stead.

Thistle's two passions were linked like the embrace of two lovers. David Stead's commitment to wildlife preservation, socialism, education and international peace became the commitments in Thistle's life also. Her book *Wild Flowers of Australia* (1938), which brought her lasting recognition, was the brainchild of David Stead. His home, Boongarre at Watson's Bay, was home for Thistle for more than fifty years, and in her old age she could not leave because he was still there, "We were one person."

For thirty years Thistle threw her time, her energy, her financial support, into David Stead's memorial, Wirrimbirra Sanctuary at Bargo. For her it was critical that this memorial have continuity and protection; her irrational plan to close the Wild Life Preservation Society, a group she had helped David Stead to foster, was merely a ruse to obtain its funds for Wirrimbirra; she sold half the land on which her house stood to enable funds to be injected into the Sanctuary; and the situation after her death when the Will she made to protect Wirrimbirra caused difficulties in securing her long-term plans for the Sanctuary — that would have caused her much distress.

Her knowledge of gardening with native plants was derivative to a certain degree; but she had the enthusiasm and drive to seek out the necessary expertise from amateurs and professionals alike and then take the initiative, especially in the early days, to present the information she acquired in an attractive and simple format for the native plant enthusiast.

"Thistle was a person with an unusual abundance of physical and mental energy," said one of her colleagues.[27] "Those who knew her were often amazed and sometimes flustered by her drive and desire to get things done. Even well into her eighties those rapid footsteps could be heard coming down the corridors, unmistakably those of TYS, approaching the meeting full of fresh ideas to be tackled immediately . . . [she] was a kind and generous person, a fact that was often disguised behind her aggressive desire to get things done. Her generosity towards the Foundation is well known, her acts of generosity to others are not so well known but were equally significant."

For many years in her role as an office-bearer for the Wild Life Preservation Society of Australia, Thistle wrote, spoke, and acted as an advocate on many issues touching wildlife and its habitats. Her aggressive attitude on many occasions did not endear her to either colleagues or adversaries, but her sincerity and tenacity were never in doubt. "Obsessive, dogmatic and somewhat eccentric," said someone who knew her well.

There was a whole generation of teachers who profited from her pragmatic approach to the teaching of science, both primary and secondary. Her uncompromising belief in the value of outdoor-education, the hands-on approach, the encounter with the natural world beyond the classroom, was handed on to aspiring teachers who took her ideas into the schools of New South Wales. "Thistle's most significant achievement I feel was the way in which she used her love of botany and Australian plants to educate people about environmental issues in the broader sense," said one of her Wirrimbirra colleagues.[28] "She managed to reach and inspire many people particularly during her years as a lecturer, and achieved this in a male dominated profession at a time when conservationists were seen as an extremist minority. We owe much to people like Thistle as they have brought about an environmental awareness that is now a part of our lives and our future."

To the general public Thistle was a charismatic figure. Her family and friends knew she could use people, that she could be stubborn and at times stupid. But one old friend summed her up, "A country needs people like Thistle. Ordinary nice people don't leave footsteps in the sands of time. People who are a bit different and prepared to walk over people are the ones who make history."

Chapter 11 Notes

1. Letter from T. Stead to K. Geeves, 20.6.82.
2. ibid.
3. C. Hinder, pers. comm. 12.11.96.
4. Letter from T. Stead to K. Geeves, 9.11.82.
5. H. Rowley, pp. 529, 543, 553, 558, 560, 561.
6. Letter from T. Stead to K. Geeves, 17.11.83.
7. J. Teulon, pers. comm. 3.6.97.
8. K. Geeves, pers. comm. 11.1.97.
9. Information supplied by Mrs Vivienne Webb.
10. Letter from T. Stead to K. Geeves, 23.9.84.
11. ibid. 14.1.85.
12. ibid. 17.8.85.
13. Letter from Vice-Chancellor, Professor K. R. McKinnon, 8.7.85.
14. Letter from T. Stead to K. Geeves, 17.8.85.
15. ibid. 29.12.85.
16. ibid. 22.5.86.
17. ibid.
18. ibid. 3.3.87.
19. ML MSS 5715 5(25).
20. Copy of letter in personal papers held by the family.
21. Letter from T. Stead to K. Geeves, 10.6.87.
22. ibid. 14.1.88.
23. ibid.

24. J. Stead, pers. comm. 14.7.97.
25. K. Geeves, pers. comm. 20.2.97.
26. J. Stead, pers. comm. 14.7.97.
27. N. H. Hicks in *Wildlife Research News*, September 1990, p. 5, 6.
28. P. Cuneo in *Wildlife Research News*, September 1990, p. 8.

Appendix

List of organizations to which Thistle Stead belonged:

- Australian Society of Authors
- Australian Conservation Foundation
- Australian Systematic Botany Society
- Institute of Landscape Architects
- Burrendong Arboretum Society
- David G. Stead Memorial Wildlife Research Foundation
- Field Naturalists Club of Victoria
- National Parks Association of New South Wales
- Teachers' Federation (New South Wales)
- Nature Conservation Council of New South Wales
- New Education Fellowship
- Public Service Association
- Society for Growing Australian Plants
- Tasmanian Conservation Trust
- Wildlife Preservation Society of Australia
- Greening of Australia
- Naturalists' Society of New South Wales
- Naturalists' Society of Western Australia
- Association for Environmental Education (New South Wales)

Bibliography

1. *Australian Wild Life,* publication of the Wildlife Preservation Society of Australia.
2. Boardman, G. *et al. Sydney Teachers College —A History 1906–1981,* Hale and Iremonger: Sydney, 1995.
3. Bradley, J. *Bringing Back the Bush — The Bradley Method of Bush Regeneration* ed by J. Larking, A. Lenning, J. Walker. Lansdowne Press: Sydney, 1988.
4. Gilbert, L. A. *The Royal Botanic Gardens, Sydney: A History 1816–1985.* Oxford University Press: Melbourne, 1986.
5. Harris, Thistle Y. *Wild Flowers of Australia,* Angus and Robertson: Sydney, 1938.
6. Harris, Thistle Y. *Australian Plants for the Garden.* Angus and Robertson: Sydney, 1953.
7. Harris, Thistle Y. *The Teaching of Nature Study.* Australian Council for Educational Research: Melbourne, 1954.
8. Harris, Thistle Y. (ed). *Naturecraft in Australia.* Angus and Robertson: Sydney, 1956.
9. Harris, Thistle Y. *Australian Plant Life.* Around Australia Program, 1966.
10. Harris, Thistle Y. *Alpine Plants of Australia.* Angus and Robertson: Sydney, 1970.
11. Harris, Thistle Y. *Gardening with Australian Plants Vol. 1 Shrubs.* Thomas Nelson (Aust.) Pty Ltd, 1977.
12. Harris, Thistle Y. *Gardening with Australian Plants Vol. 2 Small Plants and Climbers.* Thomas Nelson (Aust.) Pty Ltd, 1979.
13. Harris, Thistle Y. *Gardening with Australian Plants Vol. 3 Trees.* Thomas Nelson (Aust.) Pty Ltd, 1980.
14. Latreille, A. *The Natural Garden. Ellis Stones: His Life and Work.* Viking O'Neil: Victoria, 1990.
15. Rowley, H. Christina Stead. *Minerva Australia, 1994.*
16. *Selected Letters (of C. Stead).* Edited with preface and annotations by R. G. Geering. Angus and Robertson: Sydney, 1992.
17. Stead, C. *The Man Who Loved Children.* Angus and Robertson, Classic Edition, 1995. (First published 1941.)
18. *Wildlife Research News*, publication of David G. Stead Wildlife Research Foundation of Australia.
19. Williams, C. *Christina Stead, A Life of Letters.* McPhee Gribble: Melbourne, 1989.

Index